Law and Marxism

T0341599

Law and Marxism

A General Theory

Evgeny Pashukanis

PLUTO  PRESS

First published 1987 by Pluto Press
New Wing, Somerset House, Strand, London WC2R 1LA
and Pluto Press, Inc.
1930 Village Center Circle, 3-834, Las Vegas, NV 89134

www.plutobooks.com

Copyright © Evgeny Pashukanis 1987

The right of Evgeny Pashukanis to be identified as the author of this work has
been asserted in accordance with the Copyright, Designs and Patents Act 1988.

British Library Cataloguing in Publication Data
A catalogue record for this book is available from the British Library

ISBN 978 0 86104 740 6 paperback

Printed and bound by CPI Group (UK) Ltd, Croydon, CR0 4YY

Contents

	Page
Notes to This Edition	7
Editor's Introduction	9
Preface to the German Edition	33
Preface to the Third Russian Edition	36
Preface to the Second Russian Edition	37
Introduction: The Tasks of General Legal Theory ...	47
1: The Methods of Constructing the Concrete in the Abstract Sciences	65
2: Ideology and Law	73
3: Norm and Relation	85
4: Commodity and Subject	109
5: Law and the State	134
6: Law and Morality	151
7: Law and the Violation of Law	166
Appendix: An Assessment by Karl Korsch	189

Notes to This Edition

The translation is taken from E. Paschukanis, *Allgemeine Rechtslehre und Marxismus: Versuch einer Kritik der juristischen Grundbegriffe;* mit einer Rezension von Karl Korsch; Frankfurt a.M.; Verlag Neue Kritik, 1967, photmechanischer Nachdruck der deutschen Ausgabe von 1929, Wien und Berlin: Verlag für Literatur und Politik. (Deviations from this edition are marked in the text.) Other translations consulted were: E-B. Pasukanis, *La Théorie Générale du Droit et le Marxisme*, présenté par J.-M. Vincent, traduit par. J.-M. Brohm, Paris; EDI (Etudes et Documentation Internationales), 1970; and E. Pashukanis, 'The General Theory of Law and Marxism', edited by J. N. Hazard, translated by E. Babb, in *Soviet Legal Philosophy*, Cambridge, Mass.: Harvard University Press, 1951 (Twentieth Century Legal Philosophy Series).

We substitute for quotations from, and references to, Marx and Engels, details of current English editions, and locate, in addition, numerous such sources for which Pashukanis gave no details. Notes in the text refer to the following editions: Karl Marx, *Capital*, vol. 1, introduction by E. Mandel, translated by B. Fowkes, Pelican Marx Library, General Editor Q. Hoare, Harmondsworth, Middlesex: Penguin Books (in association with *New Left Review*), 1976; Karl Marx, *Capital*, vol. III, Moscow: Foreign Languages Publishing House, 1962; Karl Marx, *Grundrisse: Introduction to the Critique of Political Economy*, translated by M. Nicolaus, The Pelican Marx Library, General Editor Q. Hoare, Harmondsworth, Middlesex: Penguin Books (in association with *New Left Review*), 1973; Karl Marx and Friedrich Engels, *Selected Works*, 3 vols., Moscow: Progress Publishers, vol. I, 1969; vol. III 1970; Marx and Engels,

Collected Works, vols. III-VII, Moscow: Progress Publishers and London: Lawrence & Wishart, vol. III, 1975; vol. IV, 1975; vol. VI, 1976; vol. VII, 1977; Marx and Engels, *Selected Correspondence*, 2nd ed., Moscow: Progress Publishers, 1965.

It should be noted that there are differences between the Anglo-Saxon and Continental traditions of law and jurisprudence. For a Marxist interpretation, such as Pashukanis', these differences are superficial and do not affect the substance of the matter. Nevertheless, problems emerge for the translator as a consequence of discrepant terminologies. One main area of problems which may be mentioned at the outset is that *pravo, Recht* and *droit* cover a broad area that in English is usually divided between 'right' and 'law'. In almost all cases we have given 'law', but this point should be borne in mind. One consequence of it is that when the European languages wish to make the distinction, easily made in English, by the terms 'right' and 'law', circumlocutions become necessary. Thus, in Chapter 3 especially, we find a distinction made between 'subjective law' ('*subjektives Recht*') and 'objective law' (*objektives Recht*'), and between 'law in a subjective sense' and 'in an objective sense' (*Recht im subjektiven sinne* and *Recht im objektiven sinne*). Our thanks go to Piers Beirne for checking some points with the 3rd Russian edition; to Yaa Luckham and Sol Picciotto for help with legal terminology; to David Evans for many helpful stylistic suggestions; and to Keith Smith for supplying an elusive reference.

Editor's Introduction

1

Evgeny Bronislavovich Pashukanis published his important contribution towards the materialist critique of legal forms in 1924. It remains to this day the most significant Marxist work on the subject. Indeed, such has been the paucity of original work in this area that in Britain the standard reference work is even older: Karl Renner's book on *The Social Functions of Law* – a product of the Marxism of the Second International. Needless to say, Pashukanis subjects Renner's theories to severe criticism.

The present revival of interest in the theories of Pashukanis forms part of the current renaissance of Marxist debate. More particularly, it is part of a process of recovery of the heritage of Bolshevik thought repressed by the Stalinist bureaucracy and its international supporters; for example – in the field of political economy – the works of I. I. Rubin (whose approach has interesting points of contact with that of Pashukanis). The appearance of this English translation of Pashukanis provides a stimulus to the development of a far-reaching theoretical criticism of law, which is essential if a properly materialist approach is to distinguish itself from a radicalism that unconsciously remains imprisoned within a bourgeois frame of reference. The recent trenchant critique of *The Politics of the Judiciary* by J. A. G. Griffiths demonstrates the class-prejudices of judges; but the *Times Literary Supplement* (6.1.78) reviewer devotes a page to fending off even this level of criticism, maintaining that one must restrict oneself to the criticism of particular decisions in legal terms. Pashukanis' bold perspective on the revolutionary development of post-capitalist society forces criticism to go beyond sniping at 'abuses', or denouncing the current content of legal norms. The revolutionary overthrow of capitalist forms of social organisation cannot be grasped in terms of a quantitative extension of existing rights; it forces us to project a qualitative supersession of the form of law itself.

2

Pashukanis was born on the 10th February 1891 of a Lithuanian family. In 1912 he became a Bolshevik.

He was one of the leading authorities in Soviet legal science; a vice-commissar of Justice; author of a series of articles in the *Encylopaedia of State and Law* (Moscow 1925-27); author of *The General Theory of Law and Marxism* (his early masterwork which went into three editions and German and Italian translations in the twenties); editor for some years of the most important journals; and author of other books, and many articles, especially on international law. Criticism of Pashukanis began to mount in the late twenties and he found it desirable to publish an article correcting his errors in 1930. This did not prevent either his writings from being burnt, or himself from being liquidated in 1937 as a member of 'a band of wreckers' and 'Trotsky-Bukharin fascist agents' (thus Vyshinsky).[1] After his fall the usual band of time-serving philistines and eclectics took over the field. After the Twentieth Congress (1956) things developed to the point where there was a call for his rehabilitation;[2] but, although it is now recognised that he and others were unjustly condemned for sabotage, Vyshinsky's negative standpoint on his work is still endorsed.[3]

[1] *Soviet Legal Philosophy*, ed. Hazard, trans. Babb; Harvard University Press 1951; p. 315. See also Roy Medvedev. *Let History Judge*, London, Macmillan, 1972, p. 524.
[2] See *State and Law: Soviet and Yugoslav Theory*, by Ivo Lapenna, London 1964, p. 55.
[3] I. V. Pavlov writes in 1957 that 'the concept of Soviet Law as dying bourgeois law, and everything that followed from that theory, and accompanied it, was finally and definitely destroyed.' (Quoted by Lapenna p. 95n.)

3

One Marxist approach to the critique of law consists in demonstrating the conformity of the content of laws and legal institutions with the material interests of the ruling class. However, what is required in the materialist interpretation of the legal sphere is not merely an investigation of the content of legal regulations but also a materialist account of the *form* of law itself.[4] It is easy to point out that modern capitalism could not exist with strict prohibitions against usury. Similarly, it is perfectly obvious that social forces were involved in the struggle over legal limitations on the length of the working day – so graphically depicted by Marx in *Capital*. Pashukanis embarks on more subtle problems: he analyses such concepts as 'legal norm', 'legal subject', and so on, which, it seems, can be taken in abstraction from any specific content.

In accordance with the principles of historical materialism, these forms must be grounded in the sub-structure but he emphasises that this does not mean that one can dismiss the purely juridical concepts as 'ideological phantasms'. A materialist account of the specific character of legal regulation is required, which explicates it theoretically in terms of its real historical significance as a necessary expression of the economic content at a specific level of the social structure. The peculiar problematic of this form must be respected, however much the claims of the ideologists need to be demystified in the light of the material determination of legal forms by the substructure. Until the existing mode of production is overthrown, these ideological forms express the nature of social relationships with a certain validity. The task of a Marxist critique of law is not to prove that juridical concepts are consciously manipulated by bourgeois publicists in order to browbeat the workers (which is indisputable), but to show that in them – in these concepts – social reality takes on an ideological form which expresses certain objective relationships arising from

---

[4] 'We . . . laid . . . the . . . emphasis . . . on the *derivation* of political, juridical and other ideological notions . . . from basic economic facts. But in so doing we neglected the formal side – the ways and means by which these notions come about – for the sake of the content.' F. Engels' letter to Mehring, July 14, 1893 (*Selected Correspondence* of Marx and Engels – Moscow 1965 – p. 459).

the social relations of production and stands or falls with them. An ideological form cannot die out except with the social conditions which generated it. The struggle against ideology, however, helps to deprive it of the capacity to mystify the social relationships out of which it grew, and to make possible a scientific politics.

If law is not explored in terms of its internal structure, then its peculiar character will be dissolved away into some vaguer notion of social control. This is all most Marxists provide. Pashukanis complains that in place of providing a concept of law in its most complete and distinct form – and thus demonstrating its significance for a definite historical epoch – they offer a purely verbal commonplace about 'external authoritarian regulation' suitable to all epochs in the development of society. A form of social life which undergoes a process of development cannot be understood through the scholastic categories of genus and species. Like all social forms, the legal system has an historical dimension. Instead of ranging widely over the ages it is better to focus our attention where law attains its maximum degree of completeness and distinctness; that is to say, it must be analysed in the context of the appropriate social relations.

If we look at Marx's great economic work we find that he sets out to analyse the law of motion of *capitalist* society. Thus he begins his investigation, not with ratiocination about production in general, but with an analysis of definite elements: the commodity-form and value. Political economy, as a theoretical discipline employing its own specific concepts, has as its object a distinct set of social relations – not some supra-historical method of maximising scarce resources or whatever. Naturally, insofar as economics concerns itself with production and distribution it is concerned with general features of social life; however, it is quite mistaken to subsume earlier and later modes of production under the same categories – nothing but trivial tautologies can be produced that way. Furthermore, so long as value relationships are absent, it is only with difficulty that economic activity is distinguishable from the aggregate of functions constituting social life as a unitary whole. With the gradual emergence of commodity relations – and especially with the advent of the capitalist mode of production – economic life becomes a separate structure without any admixture of kinship systems, political hierarchies or whatever, and its forms may be understood in terms of a set of categories pertaining to a specific form of production, one moreover, which has attained the highest degree of determinateness and differentiation from the rest of social life.

Pashukanis believes that similar considerations are wholly applicable to the general theory of law. The fundamental juridical abstractions reflect definite social relations; so the attempt to find a definition of law which would answer to human nature or social life in general – as well as to the complex and specific modern forms – must inevitably lead to scholastic and purely verbal formulae.

Pashukanis argues that *the juridical element in the regulation of human conduct enters where the isolation and opposition of interests begins*. He goes on to tie this closely to the emergence of the commodity form in mediating material exchanges. His basic materialist strategy is to correlate commodity exchange with the time at which man becomes seen as a legal personality – the bearer of rights (as opposed to customary privileges). Furthermore, this is explicable in terms of the conceptual linkages which obtain between the sphere of commodity exchange and the form of law. The nature of the legal superstructure is a *fitting* one for this mode of production. For production to be carried on as production of commodities, suitable ways of conceiving social relations, and the relations of men to their products, have to be found, and *are* found in the form of law. Pashukanis says that the material premises of legal relations were ascertained by Marx himself in *Capital*, and that the general intimations to be found there are far more fruitful for understanding legal relations than all the bulky treatises on law.

Hegel, like so many bourgeois theorists, presents economic activity as the outcome of intercourse between property owners; and property right is derived from the necessity of the concept, i.e. of the self-determination of freedom. Marx breaks with this idealism in his analysis:

'This juridical relation, which . . . expresses itself in a contract, whether such contract be part of a developed legal system or not, is a relation between two wills, and is but the reflex of the real economic relation between the two. It is the economic relation that determines the subject matter comprised in each such juridical act.'[6]

As the product of labour takes on the commodity form and becomes a bearer of value, people acquire the quality of legal subjects with rights. While things rule people through the 'fetishism of commodities', a person is juridically dominant over things because, as an owner, he is posited as an abstract impersonal subject of rights in things. Social life in the present epoch has two distinctive and complementary features: on the one hand human relationships are mediated by the cash nexus in all its forms, prices, profits, credit-worthiness and so on, in short all those relationships where people are related in terms of things; on the other hand we have relationships where a person is defined only by contrast to a thing – that is to say as a subject freely disposing of what is his. The social bond appears simultaneously in two incoherent forms: as the abstract equivalence of commodity values, and as a person's capacity to be the abstract subject of rights.

The ideological understanding of the relation of law to the sub-structure gets things upside down insofar as perfected commodity exchange is subordinated conceptually to legal forms; from a legal point of view the capacity to engage in commodity exchange is merely one of the concrete manifestations of the general attribute of a legal capacity to act. Historically, however, it was precisely commodity exchange which furnished the idea of a subject as the abstract bearer of all possible legal claims. It is only in the conditions of commodity production that the abstract legal form is necessary – it is only there that the capacity to have a right in general is distinguished from specific claims and privileges. It is only the constant transfer of property rights in the market that creates the idea of an immobile bearer of these rights. Indeed, the abstract capacity of everyone to be a bearer of property rights makes it difficult for bourgeois thought to see anything else than subjects of rights: legal fetishism complements commodity fetishism.

The 'commodity exchange school' – as it was known –

[6] *Capital*, vol. I, ch. 2, p. 178.

dominated Soviet legal science until the mid-thirties. Stuchka, leading representative of the moderate wing[6], interpreted civil law on the basis of commodity exchange relations; but Pashukanis, representing the radical wing, went further in claiming that law in general may be so related. He appeals here to the example provided by Marx who analyses exchange in terms of the labour theory of value, albeit that the price-form of commodities extends to cover things which do not contain labour or have no economic function at all. In much the same way, Pashukanis claims that public law relations, e.g. criminal law, are an extension of forms generated by relationships between commodity owners, albeit that the contents of such public law relations are less than adequate to this form.

For Pashukanis, legal forms regulate relationships between autonomous subjects – it is the subject that is the 'cell-form' of the legal system. In bringing out the specific character of such legal regulation of behaviour, he contrasts it with technical regulation by arguing that in the latter singleness of purpose can be assumed, whereas the basic element in legal regulation is contestation – two sides defending their rights. In deliberately paradoxical fashion he says that historically law starts from a law-suit.

Pashukanis illustrates the distinction between technical and legal regulation by assigning to the former such a thing as a railway timetable and to the latter a law concerning the responsibility of the railways to the consignors of freight. Those drawing up the timetable assume that all concerned are interested in the smooth running of the service whereas those parties to the freight contract have an eye to such things as who should suffer the consequences should something get lost. Rudolf Schlesinger has argued against Pashukanis that states commonly back 'technical' regulations by Criminal Codes and gives an interesting account of precisely the experience of the Railway Courts in the USSR to prove this.[7] However, he misses the point – which is that the distinction between the two facets of the matter is not thereby abolished. Clearly Railway Courts, concerned with the culpability of a negligent engine-driver, or the Supreme Court, preventing drivers who have correctly re-

---

[6] See R. Schlesinger, *Soviet Legal Theory*, p. 205.
[7] *Ibid.*, pp. 161-164.

fused to drive unsafe engines from being convicted for sabotage, have to master the relevant technical regulations if their judgments are to be soundly based. Nonetheless, the technical regulations are designed to achieve the best possible railway service, while the codes governing the allocation of responsibility for negligence, for example, have to regulate various conflicting interests – of management, workers, and travellers. Schlesinger considers it Utopian to suppose that social organisation could ever be a purely technical matter – conflicts of interest would always occur. This may be so – but whether the legal apparatus as we know it today would persist is another question.

Pashukanis' view that law arises in order to cope with certain competing interests, and that the cell-form of the legal system is the subject asserting a claim, may be questioned because it leaves out of account state coercion. It may be said to ignore the fact that his theory does not comprehend such earlier forms as 'the King's peace'. It does not focus on the relationships of dominance and subordination found in class societies based on various property relationships.

Pashukanis argues that property attains its highest development (in the shape of unimpeded possession and alienation) only in modern society, and that this freedom of disposition may be closely related to the category of legal subject or legal person. It is only by starting here that one can go on to explain precisely why class dominance in modern society is mediated by the rule of law and the modern state. (The procedure is no odder than that of Marx who starts his exposition in *Capital* with the commodity in order to arrive later at the concept of surplus value which is the *specific* form of appropriation of surplus labour in capitalism – albeit that exploitation existed in non-commodity-producing societies.)

Pashukanis thinks that the view of law as an external regulation imposed by command of authority does not bring out the specific character of legal regulation. This does not mean that the legal superstructure does not ensure the dominance of the ruling class. However, formally, the courts act as umpires in a law-suit. This form must be recognised for what it is if a materialist analysis is to expose its class character and effectively demolish its ideological function. In analysing the rule of law we need to explain why the mechanism of constraint is dissociated from the property-owners themselves, taking the form, instead, of an impersonal mechanism of judgment isolated from everyday life. In feudal times all relationships were mediated by personal dependence and authority. The obedience of the villein to the feudal lord was the direct and immediate result of the fact that the latter had an armed force at his disposal, and his authority was an inescapable God-given fact. The dependence of the wage-labourer on the capitalist is not enforced in such an immediate fashion. Firstly, the armed force of the state is a public power standing above each individual capitalist; secondly, this impersonal power does not enforce relationships of exploitation separately, for the reason that the wage-labourer is not compelled to work for a given entrepreneur but alienates his labour-power through a free contract. Since this alienation is established formally as a relationship between two autonomous commodity owners, therefore class authority must take the form of a public authority which guarantees contracts in general but does not normally constrain the independent legal subjects to accept any particular price. If the law does intervene in this way, as it is tending to do today, then law becomes much more clearly class law – except that the bourgeoisie screams the more loudly that it is not the capitalist class that rules but 'the law' (that is to say: the authority of an objective and impartial norm). However, even in the most liberal state, the rule of law is an ideological structure that endorses and enforces class rule. For, of course, the free subjects of the theory of contract are not equal except in the context of the juridical framework which recognises alienation only in its most abstract form. For basic material reasons, such as the danger of imminent starvation, the labouring class have no option but to sell their labour. They are thus dependent as a class on the capitalists as a class (albeit that each is free to choose his exploiter) and hence are justifiably characterisable as wage-slaves. There is, therefore, the coexistence of a legal form relating 'independent and equal persons' on the one hand, and, on the other, the material reality of the rule of one class over another in the bourgeois state – but mediated, as we have seen, through the rule of law.

4

The most striking of Pashukanis' positions is his implacable opposition to any concept of 'proletarian law'. Since he treats law as an historical form which achieves fullest expression in the bourgeois epoch, and which is tied closely to the commodity form, he opposes pseudo-radicalism that talks of the overthrow of bourgeois law and its replacement by proletarian law. For Pashukanis such a line is implicitly conservative since it accepts the form of law as supra-historical and capable of infinite renewal. The transition period, when the dictatorship of the proletariat oversees the revolutionary transformation of capitalism towards communism, cannot, in any case, be regarded as if it were a particular stable social formation with its own particular form of law. As for the future – a symmetrical array such as: feudal law; bourgeois law; socialist law – neglects the whole question of the withering away of the state and law in the higher stages of socialist development. For Pashukanis the end of the forms and categories of bourgeois law by no means signifies their replacement by new proletarian ones – just as the transition to communism does not mean that new proletarian categories of value, capital, and so on, appear as the bourgeois forms die out – rather the juridical element in social relations gradually disappears.

The objection may be made that, even if economic conditions change greatly, certain crimes against the person will always exist. Pashukanis believes that to reason that courts and statutes will always be necessary on this account is to mistake structures which are derived from elsewhere for essential forms in this context. As he points out, even advanced bourgeois criminology sees that anti-social behaviour is a social problem with which the jurist is ill-equipped to grapple, burdened as he is with his concepts of 'guilt' and 'responsibility' and subtle distinctions therein. If this conviction has not yet led to the abolition of the criminal courts, this is partly because transcendence of the form of law is associated with a radical deliverance from the entire framework of bourgeois society.

As a consequence of his opposition to the idea of a special proletarian form of law Pashukanis is led to the view that throughout the transition period to socialism the legal forms retained are, in reality, *bourgeois* forms.

He is able to base himself on one of Marx's texts, *Critique of the Gotha Programme* (1875), which constituted, in fact, Marx's last important political intervention. Marx's remarks illustrate the inner connection between the form of law and the commodity form. The occasion for Marx's remarks was a reaction to the programme of the newly unified German Workers Party, which stated that 'the proceeds of labour belong undiminished with equal right to all members of society' and demanded 'a fair distribution of the proceeds of labour'. Marx seizes straight away on the pious phrase 'fair distribution' in order to reassert briefly the principle of historical materialism:

> Do not the bourgeoisie assert that the present-day distribution is fair? And is it not, in fact, the only fair distribution on the basis of the present-day mode of production? Are economic relations regulated by legal conceptions, or do not, on the contrary, legal relations arise from economic ones? Have not the socialist sectarians also the most varied notions about 'fair distribution'?[8]

Historical materialism holds that disputes about what is fair in abstraction from the economic basis of society are meaningless and irresolvable. All one can do is to point out what form of distribution corresponds to a certain mode of production and study the conditions arising in the present making for a change in the mode of production. For Marxism the presentation of socialism does not turn principally on distribution but on production.

Marx next considers the concept of equal right embodied in such post-revolutionary arrangements as that in which the same amount of labour which the individual has given society in one form, he receives back in another form.

> Here obviously the same principle prevails as that which regulates the exchange of commodities, as far as this is the exchange of equal values. Content and form are changed, because under the altered circumstances no one can give anything except his labour, and because, on the other hand, nothing can pass to the ownership of individuals except individual means of

[8] K. Marx and F. Engels, *Selected Works*, Vol. III, Moscow 1970, p. 16.

consumption. But, as far as the distribution of the latter among the individual producers is concerned, the same principle prevails as in the exchange of commodity equivalents: a given amount of labour in one form is exchanged for an equal amount of labour in another form. Hence, *equal right* here is still in principle — *bourgeois right* . . . this *equal right* is still constantly stigmatised by a bourgeois limitation. The right of the producers is *proportional* to the labour they supply; the equality consists in the fact that measurement is made with an *equal standard*, labour.

But one man is superior to another physically or mentally and so supplies more labour in the same time, or can labour for a longer time; and labour to serve as a measure, must be defined by its duration or intensity, otherwise it ceases to be a standard of measurement. This *equal* right . . . tacitly recognises unequal individual endowment and thus productive capacity as natural privileges. *It is, therefore, a right of inequality, in its content, like every right.* Right by its very nature can consist only in the application of an equal standard; but unequal individuals (and they would not be different individuals if they were not unequal) are measurable only by an equal standard in so far as they are brought under an equal point of view, are taken from one definite side only, for instance, in the present case, are regarded *only as workers* and nothing more is seen in them, everything else being ignored. Further, one worker is married, another not; one has more children than another, and so on and so forth. Thus, with an equal performance of labour, and hence with an equal share in the social consumption fund, one will in fact receive more than another, one will be richer than another, and so on. To avoid all these defects, right instead of being equal would have to be unequal.

But these defects are inevitable in the first phase of communist society as it is when it has just emerged after prolonged birth pangs from capitalist society. Right can never be higher than the economic structure of society and its cultural development conditioned thereby.[9]

Pashukanis holds that Marx here characterises as a bourgeois limitation any external application of an equal standard which, necessarily, ignores the real differencess between individuals: and that Marx is therefore stigmatising law as a bour-

[9] *Ibid.*, pp. 18-19.

geois institution.

There is a subtle dialectic here; for it is not only allegedly 'unequal' law, but *any law whatever*, that gets caught up in this problem, for by applying the same standard to individuals who differ from one another, it in effect treats them unequally.

Although Marx does not remark it, the application of the standards that guarantee 'equal right' involves also a centre of authority — even when 'exploitation' has been abolished. Lenin takes this up in his analysis of this passage in *State and Revolution*:

'If we are not to indulge in utopianism,' he says, 'we must not think that having overthrown capitalism people will at once learn to work for society without any standard of right; and indeed the abolition of capitalism does not immediately create the economic conditions for such a change.

And there is no other standard than that of 'bourgeois right'. To this extent therefore there still remains the need for a state, which while safeguarding the public ownership of the means of production, would safeguard equality in labour and equality in the distribution of products.'[10]

The eventual destiny of communist society is to pass beyond this whole complex of relationships: exchange of equivalents — equal rights — public authority. However, they are inevitable in the first phase of post-capitalist development, viz. 'socialism' as Lenin labels it, rather unhappily. Marx again:

'In a higher phase of communist society, after the enslaving subordination of the individual to the division of labour, and therewith also the antithesis between mental and physical labour, has vanished; after labour has become not only a means of life but life's prime want; after the productive forces have also increased with the all-round development of the individual, and all the springs of co-operative wealth flow more abundantly — only then can the narrow horizon of bourgeois right be crossed in its entirety and society inscribe on its banners: From each according to his ability, to each according to his needs.[11]

---

[10] V. I. Lenin, *Selected Works*, London 1969, p. 332.
[11] 'Critique of the Gotha Programme'. *Selected Works*, Vol. 3, p. 19.

This idea of a passage beyond the narrow horizon of bourgeois right needs careful scrutiny. For example, in his commentary Lenin talks, on the one hand, about this transition producing 'justice and equality' or advancing humanity from formal equality to actual equality; and on the other hand he stresses that it involves the replacement of abstract standards by direct voluntary participation in labour and the free satisfaction of needs.

> 'It will become possible for the state to wither away completely when society adopts the rule: "From each according to his ability, to each according to his needs", i.e. when people have become so accustomed to observing the fundamental rules of social intercourse and when their labour becomes so productive that they will voluntarily work according to their ability. The narrow horizon of bourgeois right, which compels one to calculate with the coldheartedness of a Shylock whether one has not worked half an hour more than someone else, whether one is not getting less pay than anyone else – this narrow horizon will then be crossed. There will then be no need for society to regulate the quantity of products to be received by each; each will take freely according to his needs.'[12]

Taking Marx's discussion as a whole it is clear that the 'rule' ('from each according to his ability, to each according to his need') is not a *prescription* (not even one prescribed to the individual by himself) issued by an appropriate authority, assigning various rights and duties, but simply a *description* of the state of affairs obtaining when labour has become 'not only a means of life but life's prime want and the springs of co-operative wealth flow more abundantly'.

Examples of principles which are applied equally to all members of society by an authority capable of enforcing them, are those cited by Lenin in his discussion of 'socialism': 'He who does not work, neither shall he eat'; and 'An equal amount of products for an equal amount of labour'.

The 'rule' of the higher phase is not such a principle enforced in order to realise 'justice and equality' even 'actual equality' – because it is not enforced at all. It is clear that both

12 Lenin, op. cit., p. 333.

'ability' and 'need' are to be determined by the possessor. Under the conditions of a 'realm of freedom' it is clearly absurd to suppose that anybody could be accused of slacking, or of being greedy; rather all expressions of individuality will be just that – expressions of free subjectivity – not obedience to an objective norm.

In this context we are not only envisaging the disappearance of public authority but also of such 'internalisations 'as 'habit' or 'conscience', because labour has become unalienated and free – 'life's prime want' as Marx puts it in the *Critique.*

It is envisaged that the material basis of society in the higher phase of communism, characterised by the features mentioned by Marx, will make possible spontaneously produced forms of social behaviour and organisation, unmediated by prescriptions enjoining justice and equality, fairness, or whatever. It would therefore be mistaken to read the rule as a rule of equality which for the first time in history gets beyond treating people from one definite side only, and instead allows for individual differences by taking people as human beings with a varied range of abilities and needs. There is no way in which such an indeterminate principle could be adjudicated in the phase of its application. No one can tell me what my abilities and needs are: only I can be the final authority on that.

People can be 'measurable by an equal standard only in so far as they are brought under an equal point of view' and yet, Marx reminds us, 'they would not be different individuals if they were not unequal'. In truth the demand for equality, or for equity in economic and legal arrangements, does not go beyond a radical bourgeois framework and does not grasp the qualitative break with previous forms that Marx looks forward to. Equality is the highest concept of bourgeois politics. It is not accidental that Marx never issued any programmatic declaration for it.[13] It would be interesting to take Pashukanis further

---

[13] 'Equal rights' in the Rules of the I.W.M.A. was imposed on Marx; he wrote to Engels (November 4, 1864): 'Only I was obliged to insert two phrases about "duty" and "right" into the preamble to the Rules, ditto about "truth, morality and justice", but these are placed in such a way that they can do no harm.' (*Selected Correspondence*, Moscow 1965, p. 148.) Engels, in his comments on the Gotha Programme, condemns 'the idea of socialism as a realm of equality' as 'a one-sided French idea' which only causes confusion (*Selected Correspondence*, p. 294).

and to work out the connections between the equality of units of abstract labour in value exchanges; equality before the law of isolated subjects capable of (property) rights; equal voting power of abstract citizens[14] in bourgeois democracy; and the common humanity posited in bourgeois ethics as inhering in everyone in virtue of which all are equally worthy of respect.[15]

All this has nothing to do with Marx's communist perspective[16] based on the social individual. The materialist account of human nature as the product of the ensemble of social relations[17] knocks on the head our 'common humanity' as an abstract essence, hypostatised in us individually, whereby each claims equality with others. A mode of social life which overcomes the estrangement manifested in the present isolation and opposition of bourgeois individuals will have no place for a concept of equality. The possibility of an immanent critique of bourgeois conditions in terms of 'equality' no doubt exists because this ideal proclaimed by the French Revolution cannot be fully realised (especially in the context of the dynamics of bourgeois property relations); but this 'political' conception is beside the point in the positive elaboration of the production and reproduction of communal life under socialism. For Marx the presentation of socialism turns on the new mode of production rather than on questions of distributive justice, and he complains about the crime of 'perverting the realistic outlook by means of ideological nonsense about rights and other trash so common among the democrats and French socialists'.[18]

In the USSR enormous differences in the standard of living of different strata of society were defended by Stalin who demagogicaly called criticism of it: 'petit-bourgeois egalitarianism'. Naturally we do not defend such abuses ourselves but we would argue that the main point is not the question of the width of differentials so much as the nature of the political

---

[14] See Marx's *On the Jewish Question.*
[15] Cf., Kant's 'end in himself'; Mill's 'each to count for one'; and a modern article on 'Equality' by Bernard Williams in *Philosophy, Politics and Society,* second series, ed. P. Laslett and W. G. Runciman (Blackwell, Oxford 1962).
[16] See Marx's *Economic and Philosophical Manuscripts* (1844) – chapter on communism and private property.
[17] Marx: Thesis 6 *On Feuerbach.*
[18] *Critique of the Gotha Programme* in *Selected Works,* Vol. 3, p. 19.

process which determines such questions. In the USSR the arrogation of power over these decisions by a bureaucratic elite naturally issued in their awarding themselves a high income. But then the USSR is a far cry from the socialism Marx presumes in his discussion.

To return to Pashukanis: he sees two things — first of all that there is a close connection between the form of law and the equal standard implicit in commodity production and exchange — secondly that there is no proletarian stage between bourgeois right and the dying out of law altogether. For law will die out 'when an end shall have been put to the form of the equivalent relationship' — a relationship stigmatised by Marx as bourgeois — whereas a society which is constrained to preserve such a relationship of equivalency between labour expenditures, and compensation therefore, preserves also the form of bourgeois law.

5

Pashukanis argues that the rule of bourgeois law is preserved during the transition to socialism even when capitalist exploitation no longer exists; there is no such thing as proletarian law, eventually law dies out together with the state. The Stalinists attacked this thesis by claiming that the proletarian dictatorship must work through law of a new type — Soviet democratic law. In 1937, Pashukanis was anathematised in an article by one P. F. Yudin, and the notorious A. Y. Vyshinsky followed close behind. Their main argument was that 'the state — an instrumentality in the hands of the dominant class — creates its law, safeguarding and protecting specifically the interests of that class. There is no law independent of the state "for the reason that law is nothing without a mechanism capable of enforcing observance of the norms of law" (Lenin)'.[19] It follows that when the proletariat smashed the old bourgeois state machine and created a new revolutionary mechanism of state authority it 'inflicted a death blow on bourgeois law'.[20]

It should be noticed that in quoting Lenin's phrase about right being nothing without a mechanism capable of enforcing it, the Stalinists always fail to give the context of this remark

[19] *Soviet Legal Philosophy*, p. 286.
[20] *Ibid.*, p. 287.

– which context is a quite extraordinary claim by Lenin, leading in an entirely different direction from that taken by Yudin and Vyshinsky. In his comments on Marx's point in the *Critique of the Gotha Programme* about the continued existence of bourgeois right in the transition period, Lenin goes much further than Marx:

> In its first phase communism cannot as yet be fully developed economically and entirely free from traces of capitalism. Hence the interesting phenomenon that communism in its first phase retains the narrow horizon of bourgeois right. Of course, bourgeois right in regard to the distribution of articles of *consumption* inevitably presupposes the existence of the *bourgeois state*, for right is nothing without an apparatus capable of enforcing the observance of the standards of right. It follows that under communism there remains for a time not only bourgeois right, but even the bourgeois state – without the bourgeoisie![21]

As Lenin indicates in his last phrase, this is a paradoxical state of affairs. In view of the fact that Lenin is above all the theorist of 'smashing the bourgeois state machine' and of 'proletarian dictatorship', this claim that under communism there remains 'the bourgeois state' seems to throw his whole theory into intolerable confusion. Only a couple of pages before, when he demanded 'the *strictest* control by society *and by the state* of the measure of labour and the measure of consumption', this was to be 'exercised not by a state of bureaucrats, but by a state of *armed workers*'. He reverts to the same formula later; thus the phrase 'there remains the bourgeois state' remains an isolated reference which is never organically connected with the main drift of his argument. Furthermore, his conclusion does not follow anyway: there is no reason why the authority which regulates the distribution of consumption goods should be the bourgeois state.

However, if Lenin can argue 'bourgeois right – hence bourgeois state', why may not the Stalinists argue with more justice 'proletarian dictatorship – hence proletarian law'? Yudin again:

> The dictatorship of the proletariat is a state of a new type, the law created by that state is law of a new type; Soviet

[21] 'State and Revolution', *Selected Works*, p. 335.

democratic law which protects the interests of each and every-one of the majority of the people: the toilers.[22]

The muddle in the reasoning of the Stalinists consists in their sliding from Lenin's formula that 'law is nothing without an apparatus capable of enforcing it' to the formula that 'the state creates its law'. It is true that the proletarian dictatorship is crucial to the transformation of society from a capitalist basis to a socialist one, and that it uses the legal form to facilitate this; however, no amount of repetition of the platitude that law is nothing without a mechanism of compulsion can establish the larger claim that the state actually creates its law. This claim is, in reality, implicitly idealist: the materialist method would rather locate the conditions which 'create' laws in the economic basis of society, for, as Marx puts it, 'right can never be higher than the economic structure of society and its cultural development conditioned thereby'.

Thus the fact that the proletarian dictatorship is the crucial 'moment' in the period of revolutionary transformation should not lead to adventurist and idealist conclusions about the omni-potence of state power. What it can accomplish at any time remains limited in extent. This is precisely because a more or less long period of transition is necessary. When Marx speaks of the persistence of a form of bourgeois right through the lower stages he is emphasising in the most dramatic way that 'right can never be higher' etc.; but it is equally important that this form is under the administration of the proletariat organised as the ruling class if the direction of change is to be towards socialism. Lenin is wrong when he assumes that to enforce 'bourgeois right' a 'bourgeois state' is required. If it *was* a bourgeois state not only the form but the content of the law would be bourgeois through and through. The anti-capitalist content of the law of the transition period is indicated by such measures as the forbidding of markets in means of production and the abolition of exploitation by private capital. These measures have the negative effect of blocking a reversion to capitalist *production*. However, until the socialist mode of pro-duction is capable of a sufficiently abundant supply of goods, we have, perforce, to put up with a bourgeois mode of distri-

[22] *Soviet Legal Philosophy*, p. 290.

bution of consumer goods and the associated legal forms.

Rudolf Schlesinger attaches some significance to the fact that the codification of Soviet law occurred at the outset of the New Economic Policy. He holds that the reversion to a free market involved in the NEP, taken together with the codification, encouraged Pashukanis and his school to identify law with bourgeoisification and to project a 'Utopian' disappearance of law when socialism finally arrived.[23] This historical conjuncture does not of itself affect Pashukanis' theoretical position, of course, but it serves to introduce the point that this position is somewhat double-edged as far as its practical implications are concerned. On the one hand, a conservative line might be taken if one holds that the economic foundations of socialism will take a long time to emerge; it would follow that such a bourgeois inheritance as the form of law could not be abolished just because it was bourgeois, if it was the necessary birthmark of the new society throughout the period of its emergence. On the other hand, if one took the view that episodes like NEP were purely ephemeral and the revolutionary process would soon overthrow such limitations one could take seriously more radical perspectives. It is a significant fact about the USSR that although NEP was succeeded by regular five-year plans the low level of development of the productive forces has meant that State ownership of the main means of production is still accompanied on the side of distribution by commodity forms (including black markets) and wage forms. It is only to be expected therefore that legal forms not dissimilar to those in bourgeois regimes figure in the modes of social control. Schlesinger totally misunderstands Pashukanis' theses when he says that 'it is hardly conceivable that the social machinery protecting honest trade in the USA against unfair competition should be described as Law, and the machinery protecting the socialist way of production in the USSR against speculation should not'.[24] Naturally the persistence of the material conditions providing a continuing temptation to speculate makes necessary a legal apparatus to deal with it – but this is a bourgeois form to crush a bourgeois vice.

However, it could be argued that such problems would be

[23] *Soviet Legal Theory*, p. 92 and p. 149.
[24] *Ibid.*, p. 159.

less persistent after revolution in an advanced capitalist country which would involve a much shorter period of transition and make relevant Pashukanis' more radical perspectives.

The importance of Pashukanis, as far as Marxist politics is concerned, is that he casts doubt on the view, common to Stalinists and Social-democrats alike, that the form of law is essentially neutral and can be filled with a given class content according to the will of the dominant class – a change in those issuing the laws is all that is necessary for progress – so the Stalinists can continue to operate law and the state 'without the bourgeoisie' into the socialist epoch, and there seems no reason why, as part of some historic compromise, 'proletarian' laws should not be established side by side with those favouring the bourgeoisie.

6

The most difficult point in Pashukanis' argument is his handling of form and content. From a dialectical point of view a form is the form of its content, and one may be alarmed at the outset if one imagines that Pashukanis proposes to write a treatise on legal forms in abstraction from content. However this would be a misunderstanding. In characterising law as a *bourgeois* form he clearly *is* relating law to a definite material content – the social relations founded on commodity exchange. This is also the basis of his confidence in the possibility of its supersession under communist conditions.

A difficulty that arises from a Marxist point of view is that the bourgeois regime is one of *generalised* commodity production; that is, it treats labour-power as a commodity and pumps out surplus labour from the wage-workers. Yet Pashukanis makes reference to commodity exchange without taking account of the various forms of production that might involve production for a market – for example the sphere of simple commodity production by self-employed craftsmen, or the slave-labour incorporated in many commodities traded in the ancient world, as well as modern capitalist production based on wage-labour. The suspicion arises that he has failed to correlate the form of law with a definite system of relations of production because reference to the level of market exchange is insufficiently precise. He does not say anything about that essential indicator of bourgeois relations – the extraction of surplus value by the class owning the means of production. Marx himself, for that matter, might be said to be mistaken in arguing that *bourgeois* right persists under forms of equivalent exchange even where exploitation based on ownership of capital is absent (and hence, *a fortiori*, Lenin would not need his 'bourgeois state without the bourgeoisie').

In my view, complaints on this score are misplaced, for it is precisely one of the interesting features of bourgeois exploitation that it inheres in economic relations that *do not* achieve formal legal expression. Formally speaking, Pashukanis is *correct* to refer law only to social relationships based on commodity exchange. Commodity exchange relations did have some weight in the Roman world, hence the possibility of modern codifications utilising Roman law; yet it is historically the case that such relations have greatest social weight in the bourgeois epoch when *generalised* commodity production allows the constitution and reproduction of capitalist domination. Pashukanis *should* perhaps have laid greater stress on the need to criticise law not only on the basis of what it shows (the fetishisation of relationships of commodity exchangers) but on what it does not, and cannot, show, and, indeed, ideologically cloaks: the inner world of capitalism's appropriation of labour-power once the latter has commodity-form.

The monopolisation of the means of production by the capitalist class is an *extra*-legal fact (quite unlike the political-economic domination of the feudal lord). The bourgeois legal order contents itself with safeguarding the right of a property owner to do as he wishes with his own property – whether it be the right of a worker to sell his labour power because that is all he owns, or that of the capitalist to purchase it and retain the product.

Marx says: 'The sphere of circulation or commodity exchange, within whose boundaries the sale and purchase of labour-power goes on, is in fact a very Eden of the innate rights of man.'[25] It is this sphere, with its exchange of equivalents by free persons, that is expressed in juridical relations. What is not expressed therein is the character of the consumption of the use-value of the labour-power acquired; the utter subordination of the labourer to the power of capital during the labour process; the extraction of the surplus; capitalist exploitation. No amount of reformist factory legislation can overcome the basic presupposition of the law: that a property freely alienated *belongs* to the purchaser, and hence that the living labour of the worker becomes, through exchange, available for exploitation by capital. Although a consequence of generalised commodity exchange, the class domination arising is not immediately juridical in character, and is, in fact, disguised by the juridical symmetry of free exchanges between property owners. Just because of this, Marx had to move from the critique of law to the critique of political economy in order to expose the roots of capitalist domination. The task left is that of tracing on this basis both the relationships that are expressed in the legal superstructure and those that it ideologically spirits away. Pashukanis has given us the most exciting contribution since Marx to this critique of law.

C. J. Arthur
January 1978

Selected Bibliography

C. J. Arthur, 'Towards a Materialist Theory of Law' in *Critique* 7, Winter 1976/77.

Lon. L. Fuller, 'Pashukanis and Vyshinskii: a study of the development of Marxist Legal Theory', *Michigan Law Review* 1949.

J. Hazard (ed.), *Soviet Legal Philosophy*, Cambridge Mass., Harvard University Press 1951.

E. Kamenka and A. Erh-Soon Tay, 'The Life and Afterlife of a Bolshevik Jurist', *Problems of Communism* January/February 1970.

E. Kamenka and A. Erh-Soon Tay, 'Beyond the French Revolution: Communist Socialism and the concept of Law', *University of Toronto Law Journal*, 21, 1971.

H. Kelsen, *The Communist Theory of Law*, London 1955.

I. Lapenna, *State and Law: Soviet and Yugoslav Theory*, London 1964.

S. Redhead, 'The Discrete Charm of Bourgeois Law: A Note on Pashukanis' in *Critique* 9, 1978.

K. Renner, *The Institutions of Private Law and their Social Functions*, ed. O. Kahn-Freund, London, Routledge 1949.

R. Schlesinger, *Soviet Legal Theory*, London, Routledge, 2nd ed. 1951.

R. Sharlet: *Pashukanis and the Commodity Exchange Theory of Law*, 1924-1930.

Z. I. Zile (ed.), *Ideas and Forces in Soviet Legal History*, Madison, Wisc., College Printing and Publishing, 1970.

Preface to the German Edition

In bourgeois society, jurisprudence has always held a special, privileged place. Not only is it first among the other social sciences, but it also leaves its mark on them.

Not for nothing did Engels call the juridical way of looking at things the classical world view of the bourgeoisie, a kind of 'secularisation of the theological', in which 'human justice takes the place of dogma and divine right, and the state takes the place of the church'.[1]

By destroying the bourgeois state and overturning property relations, the proletarian revolution created the possibility of liberation from the fetters of legal ideology. 'The workers' lack of property' – wrote Engels in the piece quoted from above – 'was matched only by their lack of illusions'.

But the experiences of the October Revolution have shown that even after the foundations of the old legal order have collapsed, after the old laws, statutes and regulations have been transformed into a heap of waste paper, old mental habits still exhibit an extrordinary tenacity. Even now, the struggle against the bourgeois legal view of the world represents a task of pressing importance for the jurists of the Soviet Republic today. Whilst in the sphere of the theory of the state, Lenin's *State and Revolution*, published as early as November 1917, already gave a consistent and consummate Marxist view, the critical work of Marxist thought in the sphere of the theory of law began much later.

Immediately following the October Revolution, we encounter an attempt to call upon the completely and utterly un-Marxist, typically petty bourgeois psychological theory of law as a

[1] Friedrich Engels, 'Juristensozialismus', in *Neue Zeit*, 1887.

rationalisation for the immediate destruction of the old machinery of justice. The revolutionary measure – not in dispute as a political measure – of destroying the old courts, constituted by the tsarist and the Kerenski regimes, and the creation of new people's courts not bound by the norms which the October Revolution had smashed, was interpreted from the point of view of a theory which regards law as the sum of psychological 'imperative-attributive experiences'. Subsequent attempts to give this theory some substance led its adherents, in particular Reisner (who died recently), to assert that there were different systems of intuitive law existing side by side, simultaneously, within the borders of the USSR: a proletarian, a peasant, and a bourgeois system. Official Soviet law was depicted as a compromise between these systems, a kind of mixture containing all three elements. It is quite obvious that this standpoint reduces the significance of the October Revolution as a proletarian revolution to nothing, and eliminates any possibility of producing an integrated evaluation of Soviet law, or of determining the criteria for such an evaluation from the point of view of its suitability or unsuitability for the way forward to socialism.

The anti-individualistic theories of those Western European jurists who represent the so-called 'socio-economic' view of law were no less influential than the psychological theory for juridical thought in the Soviet Union. In their constructions, these jurists (Duguit, Hedemann and others) reflect modern capitalism's rejection of the principle of free competition and thus of the principle of unlimited individualism and formalism. Their theories are undoubtedly interesting, and could be made the most of in the struggle for socialist planning against bourgeois-capitalist anarchy. But they can in no way provide a substitute for a revolutionary-dialectical approach to questions of law. The task of the Marxist critique was not confined to refuting the bourgeois individualistic theory of law, but also consisted of analysing the legal form itself, exposing its sociological roots, and demonstrating the relative and historically limited nature of the fundamental juridical concepts. At the same time, there was a need to raise one's voice against any attempt to blur over the fundamental contradiction between capitalism and socialism, to veil, with the help of cleverly devised 'transformations of civil law', the class nature of capitalist private property and to attach the label of a 'social function' to it.

The Soviet state does not admit any absolute and untouchable subjective private rights. But it counterposes to this fetish neither some classless principle of social solidarity, nor the bare idea of developing the productive forces, but the concrete task of constructing socialist society and destroying the last vestiges of capitalism.

This task of elaborating a revolutionary dialectical and materialist method of jurisprudence, as opposed to the metaphysical, formal-logical, or at best historical-evolutionist method of bourgeois jurisprudence, has been taken on by the section for politics and jurisprudence of the Communist Academy.

The present work, which is recommended to the attention of the German reader, is a modest contribution towards accomplishing this task.

E. B. Pashukanis
May, 1929.

Preface to the Second Russian Edition

When I submitted my book to the public, nothing was further from my mind than that there would be a need for a second edition, and in a relatively short time at that. Moreover, I am still convinced that this came about only because the work, which was meant to provide at best a stimulus and material for further discussion, was put to a use which the author had not envisaged at all, that is as teaching material. This fact in turn can be explained by the circumstance that there is very little Marxist literature on the general theory of law. How should this be otherwise, when there was doubt in Marxist circles until very recently as to whether there was such a thing as a general theory of law?

Be that as it may, the work to hand lays no claim whatsoever to the honorary title of a Marxist text-book on the general theory of law; in the first place for the simple reason that it was largely written for the clarification of the author's own ideas. This is the reason for its abstract nature, and for the compressed form of presentation, which has the air of a first draft in places; this also explains its one-sidedness, which is unavoidable when one concentrates one's attention on particular aspects of the problem which appear to be crucial. All these characteristics make the work fairly unsuitable for use as a text-book.

Nonetheless, although I am well aware of these shortcomings, I have still decided against eliminating them. The following considerations led me to make this decision. The Marxist critique of the general theory of law is still in its early stages. Definitive conclusions will not materialise overnight; they will have to be based on rigorous elaboration of every single branch of juris-

Preface to the Third Russian Edition

There are no substantive changes in this third edition of the work as compared with the second. Naturally the reason for this is not that I had nothing to add to what I had already said or felt that further elaboration and partial revision were superfluous and impossible. On the contrary: the time has now come when the ideas which were only briefly sketched in this work can and should be set out more systematically, more concretely and more fully. The last few years have not left the Marxist theory of law untouched; at this point in time there is already sufficient material for each separate legal discipline; many individual problems have already been subjected to discussion; a foundation, albeit only a provisional one, has been laid, upon which one could base a draft Marxist guide to the general theory of law.

It is precisely because I have set myself the task of writing such a text in the very near future that I have decided against any further changes in the work to hand. It is preferable that this sketch remain what it was – a first draft of a Marxist critique of the fundamental juridical concepts.

Those footnotes which appear in this edition for the first time are indicated as such.

July, 1927

prudence. However, there is still much to be done in this respect. It will no doubt suffice to point out that the Marxist critique has not even touched on such fields as that of international law as yet. The same goes for judicial procedure and, though to a lesser extent, for criminal law. In the field of the history of law, all we have is what the general Marxist theory of history has to offer on the subject. Only constitutional law and civil law in some measure represent welcome exceptions. Hence Marxism is just beginning to conquer new territory. It is natural that this should occur at first in the form of discussion and of conflict between differing viewpoints.

My work, which raises some questions of the general theory of law for discussion, serves principally to prepare the ground in this way. That is why I have elected to preserve its original character in the main, and have not tried to adapt it to the requirements to which any text-book would have to conform. All I have done is to make those additions which were necessary, occasioned in part by suggestions made in the reviews.

I think it will be useful to make at the outset some prefatory remarks about the fundamental ideas of my work.

Comrade Stuchka has quite correctly defined my approach to the general theory of law as an 'attempt to approximate the legal form to the commodity form'. As far as I could make out from the reviews, this idea was acknowledged on the whole, despite individual reservations, as successful and fruitful. That can of course be explained by the fact that I did not, after all, have to 'discover America' in this matter. There are enough elements of such an approximation in Marxist literature, above all in Marx himself. It will suffice for me to refer, in addition to the quotations from Marx cited in the work, to the chapter 'Morality and Law. Equality', in the *Anti-Dühring*. In it, Engels gives an absolutely precise formulation of the link between the principle of equality and the law of value, with the footnote that 'this derivation of the modern idea of equality from the economic condition of bourgeois society was first expounded by Marx in *Capital*'.[1] What remained, therefore, was to fuse the individual ideas thrown up by Marx and Engels into a unity, and to attempt to think through some of the conclusions arising

[1] Engels, *Anti-Dühring*, 3rd ed., Moscow: Foreign Languages Publishing House, 1962, p. 145. [Ed.]

from this. The task consisted of this alone. The basic thesis, namely that the legal subject of juridical theories is very closely related to the commodity owner, did not, after Marx, require any further substantiation.

Similarly, the next conclusion too contained nothing new. Its import is that the philosophy of law based on the category of the subject with his capacity for self-determination (for bourgeois scholarship has not as yet created any other consistent system of legal philosophy) is actually, basically, the philosophy of an economy based on the commodity, which specifies the most universal abstract conditions under which both exchange can take place according to the law of value, and exploitation can occur in the form of the 'free contract'. This view is the basis for the critique by communists of the bourgeois ideology of freedom and equality and of bourgeois formal democracy – that democracy in which the 'republic of the market' masks the 'despotism of the factory'. This view leads us to the conviction that defence of the so-called abstract foundations of the legal system is the most general form of defence of bourgeois class interests, and so forth. If, however, the Marxian analysis of the commodity form and the closely related form of the subject has been most widely applied as a weapon in the critique of bourgeois legal ideology, it has not been made use of at all in studying the legal superstructure as an objective phenomenon. This was prevented above all by the circumstance that the few Marxists who concerned themselves with legal questions undoubtedly considered the aspect of social (state) coercive regulation as the central, fundamental, and only characteristic trait of legal phenomena. It seemed as though this viewpoint alone vouchsafed the scientific, that is to say the sociological and historical, approach to the problem of law, in contrast to the ideological, purely speculative systems of legal philosophy based on the concept of the subject with its capacity for self-determination. Hence it was quite natural to think that the Marxian critique of the legal subject which arises directly from analysis of the commodity form, had nothing to do with the general theory of law, since of course external coercive regulation of the relations between commodity owners forms only an insignificant part of social regulation as a whole.

In other words, from this point of view it seemed that everything which could be learnt from Marx's conception of the 'guardian of commodities' whose 'will resides in those objects'[2] could be applied only to a relatively narrow sphere – bourgeois society's so-called *law governing intercourse*, but was totally inapplicable to the remaining branches of law (constitutional law, criminal law, and so on) and to other historical formations such as slavery, feudalism, and so on. That is to say that, on the one hand, the significance of Marxian analysis was limited to a particular sphere of law and that, on the other hand, its findings were used *only* to expose the bourgeois ideology of freedom and equality, *only* in the critique of formal democracy, but not to throw light on the fundamental characteristics of the legal superstructure as an objective phenomenon. In the process, people failed to take two things into account: first, that the principle of legal subjectivity (which we take to mean the formal principle of freedom and equality, the autonomy of the personality, and so forth) is not only an instrument of deceit and a product of the hypocrisy of the bourgeoisie, insofar as it is used to counter the proletarian struggle to abolish classes, but is at the same time a concretely effective principle which is embodied in bourgeois society from the moment it emerges from and destroys feudal-patriarchal society. Second, they failed to take into account that the victory of this principle is not only and not so much an ideological process (that is to say a process belonging entirely to the history of ideas, persuasions and so on), but rather is an actual process, making human relations into legal relations, which accompanies the development of the economy based on the commodity and on money (in Europe this means capitalist economy), and which is associated with profound, universal changes of an objective kind. These changes include: the emergence and consolidation of private property; its universal expansion to every kind of object possible, as well as to subjects; the liberation of the land and the soil from the relations of dominance and subservience; the transformation of all property into moveable property; the development and dominance of relations of liability; and, finally, the precipitation of a political authority as a separate power, functioning alongside the purely economic power of money, and the resulting more or less sharp

[2] Karl Marx, *Capital*, vol. 1, 1976 ed., p. 178. [Transl.]

differentiation between the spheres of public and private relations, public and private law.

Thus, if analysis of the commodity form reveals the concrete historical significance of the category of the subject and the bases of the abstract schema of legal ideology, then the historical process of development of the economy based on the commodity and on money, and on capitalist commodity production, goes hand in hand with these schema materialising in the concrete form of the legal superstructure. The conditions for the development of a legal superstructure with its formal statutes, courts, trials, lawyers, and so forth, are present to the same degree that human relations are constructed as relations between subjects.

It follows from this that the basic traits of bourgeois civil law are simultaneously also the characteristic traits of the legal superstructure as such. If, at earlier stages of development, equivalent exchange, in the form of indemnification and compensation for damage done, produced that most primitive legal form which we find in the so-called *leges* of the barbarians, then in future the vestiges of equivalent exchange in the sphere of distribution, which will be retained even in a socialist organisation of production (until the transition to developed communism), will compel socialist society to enclose itself within the 'narrow horizon of bourgeois law' for a time, as Marx himself foresaw. The development of the legal form, which reaches its peak in bourgeois capitalist society, takes places between these two extremes. One can also characterise this process as the disintegration of organic patriarchal relations and their replacement by legal relations, that is to say by relations between formally equal subjects. The dissolution of the patriarchal family, in which the *pater familias* was the owner of his wife's and his children's labour power, and its transformation into a contractual family in which the spouses conclude between themselves a contract of their estate, and the children (as is the case, for example, on the American farm) receive wages from the father, is one of the most typical examples of this development. The development of relations based on the commodity and on money carries this evolution still further. The sphere of circulation, which is expressed in the formula C-M-C, plays a leading part in this. Commercial law fulfils the same function in relation to civil law as civil law does with regard to all remaining branches of law, that is to say it points in the direction of development. Thus commercial law is on the one hand a particular province, of importance only to those people whose trade it is to transform the commodity into money form and vice versa; yet on the other hand it is civil law itself, in its dynamics and in its movement towards those purest models from which every trace of the organic has been eradicated, models in which the legal subject appears in its consummate form as the indispensable and unavoidable complement of the commodity.

Thus the principle of legal subjectivity and the model it implies – which appears to bourgeois jurisprudence as the *a priori* model of the human will – follows with absolute inevitability from the conditions of the economy based on the commodity and on money. The strictly empirical and technical conception of the connection between these two aspects is expressed in observations to the effect that the development of trade fosters security for property, good law-courts, a good police-force, and other such things. But if one looks into it in greater depth, it then becomes clear, not only that one or other technical arrangement of the machinery of state is based on the market, but also that there is an indissoluble internal connection between the categories of the economy based on the commodity and on money, and the legal form itself. In a society where there is money, and hence individual private labour becomes social labour only through the mediation of a universal equivalent, the conditions for a legal form with its antitheses between the subjective and the objective, between the private and the public, are already given.

Only in a society of this kind does political power have the possibility of setting itself up in opposition to purely economic power, whose most pronounced manifestation is the power of money. Simultaneously with this, the statute form also becomes possible. It is therefore unnecessary for the analysis of the fundamental definitions of law to start out from the concept of the statute and to use it as a guide, since this concept is itself of course, as an order emanating from the political power, an appurtenance of a stage of development at which the division of society into the civil and the political has already occurred and become stabilised and at which, accordingly, the fundamental aspects of the legal form have already come into being.

Marx says that

> the *establishment of the political state* and the dissolution of civil society into independent *individuals* – whose relations with one another depend on *law*, just as the relations of men in the system of estates and guilds depended on *privilege* – is accomplished by *one and the same act*.[3]

Of course it in no way follows from what has been cited above that I regard the legal form as a 'mere reflection of purest ideology'.[4] I think I have expressed myself quite clearly enough on this issue:

> Law as a form does not exist in the heads and the theories of learned jurists alone. It has a parallel, real history, which unfolds not as a set of ideas, but as a specific set of relations.[5]

At another point I speak of the juridical concepts which 'comprise a theoretical reflection of the legal system as a system of relations'. Put another way: the legal form, expressed in logical abstractions, is a product of the actual or concrete legal form (to use Comrade Stuchka's expression), of actual mediation by the relations of production. Not only did I point out that the genesis of the legal form should be sought in the relations of exchange, but I also stressed the aspect which in my view represents the most consummate manifestation of the legal form: the law-court and the judicial process.

It goes without saying that in the development of any one legal relation there are differing, more or less pronounced ideological conceptions in the minds of those involved – about themselves as subjects, their own rights and obligations, their own 'freedom' of action, the limitations of the law, and so on.

[3] Marx, 'On the Jewish Question', in Karl Marx and Friedrich Engels, *Collected Works*, vol. III, 1975, p. 167.
[4] Petr Ivanovich Stuchka, *The Revolutionary Part Played by Law and the State: A General Doctrine of Law (Revolyutsionnaya rol' prava i gosudarstva)*, 3rd ed., Moscow, 1924, Preface. [An English translation of this work is included in *Soviet Legal Philosophy*, 1951, pp. 17-69, Transl.]
[5] See below, p. 68.

However, the practical significance of legal relations certainly does not lie in these subjective states of consciousness. So long as the commodity owner is only *aware* of himself as a commodity owner, he has not yet mediated the economic relation of exchange with all its further consequences which escape his consciousness and his will. Legal mediation is accomplished only in the instant of the contract. But a concluded business agreement is no longer merely a psychological phenomenon; it is neither an 'idea', nor a 'form of consciousness' – it is an objective economic fact, an economic relation which is inextricably linked to its similarly objective legal form.

The more or less unfettered process of social production and reproduction – formally carried out in commodity-producing society through individual private legal transactions – is the practical purpose of legal mediation. This purpose cannot be achieved with the help of forms of consciousness alone, that is to say through purely subjective aspects: it requires exact criteria, statutes, interpretation of the statute, casuistry, law-courts, and the compulsory execution of court decisions. For this reason alone one cannot limit oneself, when analysing the legal form, to 'pure ideology', nor can one disregard the whole of this objectively existing machinery. Every legal action, for example the outcome of a lawsuit, is an objective fact which has its place outside the consciousness of the parties to it in just the same way as the economic phenomenon which it mediates.

Another of the things with which Comrade Stuchka reproaches me – namely that I recognise the existence of law only in bourgeois society – I grant, with certain reservations. I have indeed maintained, and still do maintain, that the relations between commodity producers generate the most highly developed, most universal, and most consummate legal mediation, and hence that every general theory of law, and every 'pure jurisprudence', is a one-sided description, abstracted from all other conditions, of the relations between people who appear in the market as commodity owners. But a developed and consummate form does not of course exclude undeveloped and rudimentary forms, rather to the contrary, it presupposes them.

This is the way it is, for instance, with private property: only the aspect of free alienation fully reveals the fundamental nature of this institution, although property as appropriation undoubtedly existed earlier than, not just the developed form, but also the most embryonic forms of exchange. Property as appropriation is the natural consequence of every mode of production; but only within a particular social formation does property take on its logically simplest and most universal form as private property, determined by the simple precondition of the uninterrupted circulation of value according to the formula C-M-C.

Exactly the same is true of the relation of exploitation. This relation is of course in no way bound to the exchange relation, and is conceivable in a natural economy as well. But only in bourgeois capitalist society, where the proletarian figures as a subject disposing of his labour power as a commodity, is the economic relation of exploitation mediated legally, in the form of a contract. This is linked precisely with the fact that in bourgeois society, in contrast to societies based on slavery or serfdom, the legal form attains universal significance, legal ideology becomes the ideology par excellence, and defending the class interest of the exploiters appears with ever increasing success as the defence of the abstract principle of legal subjectivity.

In a word: the purport of my analysis was not at all to deny the Marxist theory of law access to those historical periods which were as yet unfamiliar with developed capitalist commodity production. On the contrary, I was and still am concerned to facilitate understanding of the embryonic forms we find in those epochs, and to link them to the more developed forms through a general line of development. The future will tell the extent to which my approach has been fruitful.

Obviously, I could only sketch the basic traits of the historical and dialectical development of the legal form in my short outline. In the process, I made use, in the main, of ideas I found in Marx. It was not my task to solve all, or even only a few, of the problems of the theory of law. I merely wished to show how one could approach them, and how the questions should be put. I am gratified by the fact alone that some Marxists found my approach to questions of law interesting and not without prospects. This further strengthens my desire to extend my work in the same direction.

Introduction: The Tasks of General Legal Theory

The general theory of law may be defined as the development of the most fundamental and abstract juridical concepts, such as 'legal norm', 'legal relation', 'legal subject' and so on. Since these concepts are abstract, they are applicable to each and every branch of law; regardless of the concrete content to which they are applied, their logical and systematic meaning remains constant. No-one would dispute the fact that the concept of the subject in civil law or in international law is subordinate to the more general concept of the legal subject as such, so that the latter category may be defined and developed independently of either concrete content. Alternatively, if we confine ourselves to one particular branch of law, we can still establish that the fundamental juridical categories cited above are not dependent on the concrete content of its legal norms, in the sense that they retain their meaning irrespective of any change in this concrete material content.

Obviously these most abstract and simple juridical concepts result from a logical elaboration of the norms of positive law and, as compared with spontaneously arising legal relations and the norms which express these, they represent the late-ripening fruit of a conscious process of creation.

Nevertheless, this does not prevent the philosophers of the Neo-Kantian school from regarding the fundamental juridical categories as prior to experience, rendering experience itself possible. Thus we read in Savalsky, for example:

Subject, object, relation and law of relation are the *a priori* of legal experience, its immutable logical premises.[1]

[1] Savalsky. *The Bases of the Philosophy of Law in Scientific Idealism* (*Osnovy filosofii prava v nauchnom idealizme*), Moscow, 1908, p. 216.

And further:

> The legal relation is the only ineluctable condition of all legal
> institutions and thus, it follows, of jurisprudence too, for with-
> out the legal relation there can be no related science, that is
> no jurisprudence, just as nature, and hence natural science,
> cannot subsist without the principle of causality.[2]

In his observations Savalsky is merely re-stating the conclusions
reached by Cohen, one of the most prominent Neo-Kantians.[3]
We find the same viewpoint propounded by Stammler in his
earlier fundamental work, *Economy and Law*,[4] as well as in his
last work, *Textbook on Legal Philosophy*. In the latter we read:

> Amongst the concepts of law one must differentiate between
> pure and qualified juridical concepts. The former are the
> general formulations of the fundamental principles of law. For
> them to be applied, nothing more is required than the notion
> of law itself. They therefore have application across the board
> to all legal questions which could conceivably arise, for they
> are nothing other than various manifestations of the formal
> concepts of law. For this reason, they should be derived from
> the fixed definitions of that concept.[5]

However fervently the Neo-Kantians may assure us that in
their view the 'idea of law' precedes experience not genetically,
in time, but logically and epistemologically, we must insist that
so-called critical philosophy leads us on this, as on many other
points, back to medieval scholasticism.

It can thus be taken as axiomatic that developed juridical
thought cannot do without a certain number of the most
abstract general definitions, irrespective of the subject with
which it is dealing. Nor can Soviet jurisprudence do without
such abstract definitions if it is to remain a jurisprudence, that
is, if it is to be equal to its immediate practical task. The fundamental, or formal, juridical concepts have a continued existence

---

[2] Ibid., p. 218.
[3] Hermann Cohen, *Die Ethik des reinen Willens* (1904), 2nd ed., Berlin,
1907, pp. 227ff.
[4] Rudolf Stammler, *Wirtschaft und Recht nach der materialistischen
Geschichtsauffassung: Eine sozialphilosophische Untersuchung*, Leipzig,
1896.
[5] Stammler, *Lehrbuch der Rechtsphilosophie* (Berlin and Leipzig, 1922),
3rd ed., 1928, p. 250.

in our statute-books and the corresponding commentaries. Also still with us is the method of juridical thought with the procedures peculiar to it.

But does this prove that the scientific theory of law should be concerned with the analysis of the above-mentioned abstractions? There is a fairly widely held view that these most general juridical concepts have a limited and purely technical relevance. We are told that dogmatic jurisprudence uses these terms for reasons of convenience alone. They have, supposedly, no further significance for theory and epistemology. Yet the fact that dogmatic jurisprudence is a practical, and in a certain sense technical, discipline does not warrant the conclusion that the concepts of this jurisprudence could not form part of a related theoretical discipline. We may agree with Karner (Renner)[6] when he asserts that the science of law begins where jurisprudence ends. It does not, however, follow from this that legal science should simply throw overboard the basic abstractions which give expression to the fundamental essence of the legal form. After all, political economy, too, began its development with practical questions, mainly in the sphere of the circulation of money. It also first set itself the task of establishing 'the means of enriching governments and the people'.[7] Nonetheless, even in these technical deliberations one can see the foundations of those concepts which have been absorbed, in a deeper and more general form, into the body of the theoretical discipline of political economy.

Can jurisprudence be developed into a general theory of law without disintegrating into either psychology or sociology in the process?

Is it possible to analyse the fundamental definitions of the legal form in the same way as political economy analyses the basic, most general definitions of the commodity-form or the

---

[6] Josef Karner, 'Die soziale Funktion der Rechtsinstitute, besonders des Eigentums', Chapter 1, p. 72, in *Marx-Studien*, vol. 1, 1904 (Karner is a pseudonym for Karl Renner). Cf. Karl Renner, *The Institutions of Private Law and their Social Functions*, edited by O. Kahn-Freund, translated by A. Schwarzschild, London, Routledge & Kegan Paul Ltd., 1949, Chapter 1, pp. 54-55.
[7] Pashukanis is here paraphrasing Adam Smith, *The Wealth of Nations*, edited by R. H. Campbell, and A. S. Skinner, Oxford: Clarendon Press, 1976, vol. I, Book 4, p. 428. [Ed.]

value-form? Whether or not the general theory of law can be regarded as an independent theoretical discipline is dependent upon these questions being resolved.

Bourgeois legal philosophy, whose representatives, in the main, are of a Neo-Kantian persuasion, resolves the problem cited above by a simple opposition of two categories (*Gesetzmässigkeiten*), the category of *Is* (*das Seiende*), and the category of *Ought* (*das Seinsollende*). In accordance with this, they postulate two very different scientific viewpoints, the explicative and the normative.

> The former approach is concerned with things as they actually are, which it attempts to render more intelligible by linking, either their inner similarities, or those congruities suggested by their external characteristics. The latter approach considers objects in the light of fixed laws, which are expressed through them, and simultaneously applies these laws as requirements to each individual object. In the former view, therefore, all facts as such are taken to have equal validity; while in the latter approach they are consistently subjected to an evaluation, either by abstracting from whatever conflicts with the laws which have been postulated, or by contrasting it, as behaviour outside the norm, to normal behaviour which substantiates the law.[8]

In Simmel, the category of *Ought* determines a particular way of thinking, which is separated by an unbridgeable abyss from that logical order wherein we conceive of the *Is* running its course with natural necessity. The concrete 'thou shalt' ('*du sollst*') can be substantiated only by reference to a further imperative. It is not possible, within the bounds of logic, to infer the imperative from necessity, nor vice versa.[9] In his most important work, *Economy and Law*, Stammler plays innumerable variations on this theme, namely the notion that the order of things (*Gesetzmässigkeit*) can be determined by two different methods: the causal and the teleological.[10] Thus, jurisprudence

---

[8] Wilhelm Wundt, *Ethik: Eine Untersuchung der Tatsachen und Gesetze des sittlichen Lebens* (Stuttgart, 1886), 3rd revised ed. in 2 vols., Stuttgart, 1903, p. 1.
[9] Cf. Georg Simmel, *Einleitung in die Moralwissenschaft* (2 vols., Berlin, 1892), Stuttgart, 1910.
[10] Stammler, *Wirtschaft und Recht*, op. cit.

would seem to have acquired a firm methodological basis as one of the normative disciplines. Indeed, the attempts to give this methodology some depth led Kelsen, for one, to the conviction that jurisprudence especially is an essentially normative science, since it may be restricted more readily than any other discipline of its kind to the confines of the formal, logical sense of the category of *Ought*. It is true that the normative is saturated with psychological elements, both in Ethics and in Aesthetics, so that it can be considered as qualified volition, that is, as a given fact, as an *Is*: the standpoint of causality continually obtrudes upon, and detracts from the purity of the normative interpretation. As opposed to this, the imperatival principle occurs in law – whose highest expression, according to Kelsen, is the law of the state – in a categorically heteronomous[11] form, which has definitively broken with the factual,

---

[11] 'Heteronomy': the previous discussion, in neo-Kantian terms, of a supposed antithesis between the viewpoint of causal connection, and that of the purely normative, would lead one to expect here not 'heteronomous', but 'non-heteronomous'. With reference to moral imperatives Kant writes: 'He can consider himself *first* – so far as he belongs to the sensible world – to be under the laws of nature (heteronomy); and *secondly* – so far as he belongs to the intelligible world – to be under laws which, being independent of nature, are not empirical but have their ground in reason alone.' (*The Moral Law or Kant's Groundwork of the Metaphysics of Morals*, edited and translated by H. J. Paton, 3rd ed., London: Hutchinson's University Library, 1956, p. 120.)

However it should be noted that Kelsen on occasion distinguishes law from morality with reference to the sanction involved; and uses the terminology of 'heteronomy' to do so: 'The more appropriate . . . the imperative form may be for the norms of an autonomous morality, the more inadequate it appears for the heteronomous rule of law.' (Quoted by O. Weinberger in his introduction to *Essays in Legal and Moral Philosophy* by Hans Kelsen, Holland, Dordrecht: D. Reidal, 1973, p. xx. See Weinberger's discussion in section 4 of his introduction.)

In another place Kelsen distinguishes two kinds of secondary norms. In the example of contract the obligation arises out of the debtor's autonomous will; whereas this is not the case with a taxation order. He goes on: 'It is this antagonism between autonomy and heteronomy which is the ground for the distinction between private and public law, insofar as this opposition is interpreted to mean that private law regulates the relations between equal subjects, while public law regulates those between an inferior and a superior subject.' (Kelsen, *General Theory of Law and State*, New York: Russell and Russell, 1961, p. 205.)

Renner says legal imperatives confront 'the individual will (autonomy)' as an 'extraneous will (heteronomy)'. (Renner, *The Institutions of Private Law*, op. cit., p. 47.) Pashukanis probably has this last connection in mind here. [Ed.]

with that which is. All that remains is to transfer the legislative function itself to the meta-juridical sphere – as Kelsen does – and all that jurisprudence has left is the pure sphere of the normative: its task is confined exclusively to ordering the various normative contents in a logically determined manner.

Unquestionably one must give Kelsen credit for one great service. As a result of his undaunted consistency, he reduced Neo-Kantian methodology, with its two categories, to absurdity. For it turns out that the 'pure' category of *Ought*, cleansed of all impurities from the *Is*, or the factual, and of all psychological and sociological 'dross', neither has, nor possibly can have, any rational definition whatsoever. Purpose itself is immaterial – a matter of indifference – to the purely juridical, that is, to the unconditionally heteronomous *Ought*. According to Kelsen, even the formulation: 'thou shalt, in order that . . .' is no longer the same as the juridical 'thou shalt'.[12]

On the plane of the juridical *Ought*, there is nothing but a transition from one norm to another on the rungs of a hierarchical ladder, at the top of which is the all-embracing, supreme norm-setting authority – a delimiting concept (*Grenzbegriff*), from which jurisprudence proceeds as from something given. One of Kelsen's critics has illustrated this attitude to the tasks of jurisprudence in the form of a caricatured conversation between a jurist and a legislator, as follows:

> 'We neither know nor care what kind of laws you should make. That appertains to the art of legislation, which is foreign to us. Pass laws as you wish. Once you have done so, we shall explain to you in Latin what kind of a law you have passed.[13]

Such a general theory of law explains nothing, and turns its back from the outset on the facts of reality, that is of social life, busying itself with norms without being in the least interested in their origin (a meta-juridical question!), or in their relationship to any material matters. Surely this can lay claim to the title of theory at best only in the sense that it is com-

---

[12] See previous note. Kant sharply distinguished between hypothetical imperatives ('thou shalt, in order that . . .') and categorical imperatives ('thou shalt . . .'), and associated morality only with the latter. [Ed.]
[13] Julius Ofner, Daz, *soziale Rechtsdenken*, Stuttgart and Gotha, 1923, p. 54.

mon practice to speak of a theory of chess. A theory of this nature has nothing to do with science. This 'theory' makes not the slightest attempt to analyse law, the legal form, as a historical form, for it has absolutely no intention of fathoming reality. For this reason it is, to put it bluntly, a waste of time.

The so-called sociological and psychological theories of law are different. One is entitled to expect more of these theories, since, by means of the method they use, they undertake to interpret law as a real phenomenon in its origin and development. But here a new disappointment awaits us. The sociological and psychological theories of law usually exclude the legal form as such from their field of observation, in other words they simply overlook the problem involved in it. They operate from the outset with concepts of a non-juridical nature, and even when they do occasionally take purely juridical definitions into consideration, they do so merely in order to label them as 'fictions', 'ideological illusions', 'projections' and so on. This naturalistic, or nihilistic attitude has a certain appeal at first sight, particularly when contrasted with the idealist theories of law which are saturated through and through with teleology and 'moralism'. After the high-sounding phrases about the 'eternal idea of law', or the 'absolute significance of the personality', the reader who is looking for a materialist explanation of social phenomena will turn with particular satisfaction to theories which treat law as the product of conflict of interest, as the manifestation of state coercion, or even as a process which takes place in the actual human psyche. Many Marxists assumed that by simply adding in the element of class struggle to the above-mentioned theories, they would attain a genuinely materialist, Marxist theory of law. Yet all that follows from this is a history of economic systems wtih a fairly faint juridical tinge, or a history of institutions, but by no means a history of law.[14]

[14] Petr Ivanovich Stuchka, *The Revolutionary Part Played by Law and the State: A General Doctrine of Law, Revolyutsionnaya rol' prava i gosudarstva*, 3rd ed., Moscow, 1924; (an English translation of this work is included in: *Soviet Legal Philosophy*, 1951, pp. 17-69; [Transl.]). Even this book, while it deals with a whole series of questions relating to the general theory of law, does not forge them into a systematic whole. In Stuchka's portrayal, the historical development of legal regulation from the point of view of its class content takes priority over the logical

While those bourgeois jurists (such as Gumplowicz) who were trying to champion materialist views in some degree, felt bound to examine closely, as it were *ex officio*, the arsenal of fundamental juridical concepts, even if only to declare them to be artificial, formal constructs, Marxists, on the other hand, who have no responsibility towards jurisprudence, usually pass over the formal definitions of general legal theory in silence, and devote their undivided attention to the concrete content of the legal norms and the historical development of legal institutions. Indeed, it needs saying at this point that when Marxist authors speak of juridical concepts, they are usually thinking of the legal regulation specific to any one epoch, in other words, they are thinking of that which people at a given stage of development look upon as law. This is evident, for example, in the following formulation:

> On the basis of a particular state of the productive forces there come into existence certain relations of production, which receive their ideal expression in the legal notions of men and in more or less 'abstract rules', in unwritten customs and written laws.[16]

Here the concept of law is examined exclusively from the point of view of its content; the question of the legal form as such is not raised at all. Nevertheless, there is no doubt but that Marxist theory should not only analyse the material content of legal regulation in different historical epochs, but should also provide a materialist interpretation of legal regulation as a specific historical form.

and dialectical development of the form itself. (The reader's attention must, however, be drawn to the fact that, if one compares the third edition of this work with the first, one cannot but notice that the author has given questions related to the legal form far greater attention in the later edition). This followed, moreover, from Stuchka's point of departure, from his conception of law as primarily a system of the relations of production and exchange. If law is seen from the outset as the form of any and all social relations, one can predict with certainty that its specific characteristics will be disregarded. Contrary to this, under fairly close scrutiny law as the form of the relations of production and exchange readily reveals its specific traits.

[16] G. Plekhanov (N. Beltov), *The Development of the Monist View of History*, Moscow: Progress Publishers, 1956, Chapter 5, p. 158. (Beltov is a pseudonym for Plekhanov.)

If, however, we forgo an analysis of the fundamental juridical concepts, all we get is a theory which explains the emergence of legal regulation from the material needs of society, and thus provides an explanation of the fact that legal norms conform to the material interests of particular social classes. Yet legal regulation itself has still not been analysed as a form, despite the wealth of historical content with which we imbue it. Instead of being able to avail ourselves of an abundance of internal structures and interconnections of the juridical, we are forced to make do with its bare outlines, only approximately indicated. These outlines are so blurred that the borderline between the sphere of the juridical and adjacent spheres is completely obliterated.[16]

It must be acknowledged that it is legitimate, up to a point, to proceed in this manner. Economic history can be described without any mention whatsoever of, for instance, the theory of rent, or wage theory.

But what would we think of an economic historian who allowed the fundamental categories of political economy – such as value, capital, rent, profit – to be submerged in the vague, undifferentiated concept of economics? Not to speak of the reception with which any attempt to pass off such economic history as a theory of political economy would be greeted. Nonetheless, this is exactly how matters stand at present with Marxist legal theory. To be sure, there is some comfort in the fact that the jurists are still searching in vain for a definition of their concept of law. Even if most lectures on general legal theory usually begin with some formula or other, actually this formula provides at best only an unclear, unarticulated approximation of the juridical. It can be postulated as axiomatic that we learn very little indeed about what law really is from the definitions it is given and, conversely, that the less an academic clings to his own definition of it, the more likely he is to succeed in making us familiar with law as a form.

[16] Mikhail Nikolayevich Pokrovsky's book, *Contributions to the History of Russian Culture (Ocherki po istorii ruskoy kultury)*, 2nd ed., Moscow, 1918, vol. I, p. 16), provides us with an example of the way in which a wealth of historical detail is reconciled with the most cursory sketch of the legal form. In it, the definition of law is reduced to the characteristic of its inflexibility and inertia, in contrast to the mobility of economic phenomena.

The cause of this state of affairs is quite obvious: a concept as complex as that of law cannot be fully comprehended within a definition formulated according to the rules of scholastic logic *per genus et per differentiam specificam*. Regrettably, those few Marxists who are concerned with the theory of law have fallen prey to the temptations of scholastic wisdom. Thus Renner, for example, bases his definition of law on the concept of an imperative addressed to the individual by society.[17] He assumes that this somewhat unimaginative construct is adequate to the task of exploring the past, present and future of legal institutions.[18]

The fundamental shortcoming of such formulae is their inability to comprehend the concept of law in its actual workings, which expose the whole wealth of its internal connections and correlations. Instead of presenting us with the concept of law in its most distinct and consummate form, thereby demonstrating its relevance to a particular historical epoch, they serve up an empty platitude about 'external authoritarian regulation', which applies equally to all epochs and all stages of social development. There is an exact analogy in political

[17] Karner (Renner), *The Institutions of Private Law*, ed. cit., Chapter 1, section i, pp. 45-48.

[18] Cf. also in N. I. Ziber, *Collected Works* (*Sobranie sochineniya*), vol. II, p. 134: 'Law is nothing other than the aggregate of the coercive norms which provide a typical example of the course taken by economic phenomena, a form whose task it is to guard against and suppress aberrations from the ordinary run of things.' We find analogous definitions of law as a coercive norm decreed by the state authority in Nikolay Bukharin's *Historical Materialism: A System of Sociology*, Ann Arbor, Michigan: University of Michigan Press, 1969, p. 157. Bukharin differs from Ziber, and especially from Renner, in placing particular emphasis on the class nature of state power, and hence of law. Podvolotsky, one of Bukharin's pupils, gives a detailed definition: 'Law is a system of coercive social norms reflecting the economic and other social relations of any given society, which are introduced and maintained by the state authority of the ruling class in order to sanction, regulate and consolidate these relations, thereby consolidating the rule of that class'. (Podvolotsky, *The Marxist Theory of Law, Marksistskaya teoriya prava*, 2nd ed., Moscow, 1926). All these definitions emphasise the relationship between the concrete content of legal regulation and economic matters. Yet at the same time they attempt to give a rigorous definition of law as a form by distinguishing external, state-organised coercion; basically, that is, they do not transcend the crudely empirical method of that very practical or dogmatic jurisprudence whose overthrow should be the task of Marxism.

economy, in the attempts to find a definition of the concept of economics which would encompass all historical epochs. If the whole of economic theory consisted of such sterile scholastic generalisations, it would scarcely merit being called a science.

As we well know, Marx begins his investigation, not with observations about the economy in general, but with an analysis of the commodity and of value. For the economy only begins to be differentiated as a distinct sphere of relations with the emergence of exchange. So long as there are no relations determined by value, it is difficult to distinguish economic activity from the remaining totality of life's activities together with which it forms a synthetic whole. Pure natural economy cannot form the subject of political economy as an autonomous science.[19] The relations of capitalist commodity production alone form the subject-matter of political economy as an independent theoretical discipline employing its own specific concepts.

Political economy begins with commodities, begins from the moment when products are exchanged for one another – whether by individuals or by primitive communities.[20]

The same applies without reservation to the general theory of law. Those basic juridical abstractions representing the closest approximations to the legal form as such, which are generated in the course of the development of juridical thought, reflect quite specific, very complex social relations. Any attempt to discover a definition of law corresponding, not only to these complex relations, but also to 'human nature', or to the 'human commonwealth', must inevitably result in empty scholastic verbal formulae.

[19] It must furthermore be added that, among Marxists, complete unanimity does not prevail as to what constitutes the subject-matter of theoretical economics. This is amply demonstrated in the discussion of Ivan Ivanovich Stepanov-Skvortsov's article in *Vestnik Kommunisticheskoy Akademii*, 1925, no. 12. Stepanov's view that the categories of commodity-production and capitalist commodity-production (which I have mentioned) in no way form the specific subject of theoretical economics was nonetheless rejected out of hand by the overwhelming majority of our [Soviet] political economists who participated in the discussion.
[20] Friedrich Engels, 'Review of the Critique of Political Economy', in Karl Marx and Friedrich Engels, *Selected Works*, vol. I, 1969, p. 514.

58

When one is then forced to turn from this lifeless formula to the analysis of the legal form as it actually occurs, one encounters a series of obstacles. These obstacles can be overcome only by way of notorious stratagems. Hence we are told – but usually only after having first been given a general definition of law – that there are actually two kinds of law: subjective and objective law, *jus agendi* and *norma agendi*. The problem with this is that the definition itself in no way admits of such a dichotomy, so that one is forced either to negate one of these two kinds of law, presenting it as fictitious, illusory and so forth, or to assume a purely superficial connection between the general concept of law and its two forms. Nonetheless, this dual nature of law, its split into norm and legal power,[21] is of similarly vital significance as is, for example, the split of the commodity into use value and exchange value.

Law as a form is comprehensible only within its most precise definitions. It exits only in antitheses: objective law – subjective law; public law – private law, and so on. Yet all these fundamental distinctions will turn out to be mechanistically appended to the basic formula if we expect this to span all epochs and stages of social development, even including those which knew absolutely nothing about the antitheses cited above.

It is only with the advent of bourgeois-capitalist society that all the necessary conditions are created for the juridical factor to attain complete distinctness in social relations.

Leaving aside altogether the culture of primitive peoples, in which it is difficult to distinguish law from the total mass of normative social phenomena, one finds that even in medieval Europe only embryonic legal forms existed. All the antitheses mentioned above fuse into an undifferentiated whole. There is no clear dividing line between law as objective norm and law as legal power.[22] They make no distinction between the general

---

[21] '*Norm und Rechtsbefugnis*': See following note. [Ed.]

[22] '*Recht als objektive Norm und Recht als Berechtigung*': (By 'right' (*Berechtigung*) is understood not the mere reflex right, but the legal power to assert (by taking a legal action) the fulfilment of a legal obligation, that is, the legal power to participate in the creation of a judicial decision constituting an individual norm by which the execution of a sanction as a reaction against the non-fulfilment of an obligation is ordered.) (Kelsen, *Pure Theory of Law*, translated by M. Knight, Berkeley and Los Angeles: University of California Press, 1967, p. 168.) [Ed.]

norm and its concrete application; as a result the spheres of activity of judge and legislator are indistinguishable. Both in the Mark community[23] and in the organisation of the feudal system, the antithesis between public and private law is obliterated. The opposition of man as a private person to man as member of a political grouping – so typical of the bourgeois epoch – is completely absent. A long process of development was necessary for all these facets of the legal form to become crystallised with complete precision. The cities were the most important stage upon which this process was acted out.

Therefore the dialectical development of the fundamental juridical concepts not only provides us with the legal form as a fully developed and articulated structure, but also reflects the actual process of historical development, a process which is synonymous with the process of development of bourgeois society itself. The objection that, as a discipline, the general theory of law deals only with formal, qualified definitions and artificial constructs, cannot be raised against our interpretation of it.

There is no doubt that political economy studies something which actually exists, even though Marx himself was the first to point out that, in the discovery of entities like value, capital, profit, rent and so forth, 'neither microscopes nor chemical reagents are of assistance'.[24] The theory of law makes use of abstractions which are no less 'artificial': the research methods of natural science cannot discover a 'legal relation', or a 'legal subject' either, yet behind these abstractions too lie perfectly real social forces.

To a person living in a natural economy, the economics of value relations would appear as an artificial distortion of simple, natural things, in just the same way as the legal turn of mind appears to run counter to the 'common sense' of the ordinary person.

It is worth noting that the 'ordinary person' is mentally far

[23] 'Markgenossenschaft': a social organisation, resting on communal ownership of land by small groups of freemen of which traces are supposed to remain in the land tenures of England and Germany. Engels wrote an article on 'the Mark' as an appendix to an edition of Socialism, Utopian and Scientific: 'Die Mark' in Entwicklung des Sozialismus von der Utopie zur Wissenschaft, Zürich: Hottingen, 1882. [Ed.]

[24] Karl Marx, Capital, vol. I, Preface to 1st ed., 1976 ed., p. 90. [Ed.]

less accustomed to the legal than to the economic viewpoint. For even in a situation where the economic relation simultaneously materialises as a legal relation, in the overwhelming majority of cases it is the economic aspect of it which really interests the participants, whilst the juridical aspect remains in the background, becoming fully exposed to view only in exceptional cases (trial, litigation). Further, the people who personify the 'juridical aspect' in its sphere of activity are generally members of a particular caste (lawyers, judges). That is why ordinary people are more used to thinking in economic categories, and find this more natural than thinking in juridical categories.

To think that the juridical concepts which express the sense of the legal form have been quite arbitrarily dreamt up is to fall into the error which Marx pointed to in the work of the Enlightenment scholars of the eighteenth century. According to Marx, these scholars were still unable to account for the origin and development of the enigmatic forms assumed by human relations, so they tried to render these forms less incomprehensible by declaring them to be in fact human inventions rather than something which appeared out of the blue.[25]

However, it cannot be denied that a large number of juridical constructs (most of the constructs of public law, for example) are really extraordinarily arbitrary and questionable. We shall try, in what follows, to elucidate the reasons for this. For the time being, we shall confine ourselves to the observation that the value form becomes universal under the conditions of developed commodity production and assumes, besides its primary forms, various derived and artificial forms of expression. Thus it appears as the price of things which are not the products of labour (such as land), or even of things which have absolutely nothing to do with the production process (for instance, military secrets purchased by a spy). Nevertheless, this does not alter the fact that value as an economic category can be comprehended only from the standpoint of the socially necessary expenditure of labour required in the production of a given commodity. In the same way, the universality of the legal form does not necessarily call a halt to our search for those relations on which it is actually based. Later, we hope to be able to establish that this basis is *not* synony-

[25] Ibid., pp. 185-186.

mous with those relations known as public-law relations.

A further objection to our conception of the tasks of general legal theory is that the underlying abstractions are regarded as exclusive to bourgeois law. Proletarian law – we are told – should find alternative general concepts, and the search for such alternatives should be the task of the Marxist theory of law.

At first sight, this objection appears to be a very serious one. Yet it rests on a misunderstanding. In raising a demand for new general concepts specific to proletarian law, this line appears to be revolutionary par excellence. In reality, however, this tendency proclaims the immortality of the legal form, in that it strives to wrench this form from the particular historical conditions which had helped bring it to full fruition, and to present it as capable of permanent renewal. The withering away of certain categories of bourgeois law (the categories as such, not this or that precept) in no way implies their replacement by new categories of proletarian law, just as the withering away of the categories of value, capital, profit and so forth in the transition to fully-developed socialism will not mean the emergence of new proletarian categories of value, capital and so on.

The withering away of the categories of bourgeois law will, under these conditions, mean the withering away of law altogether, that is to say the disappearance of the juridical factor from social relations.

The transition period – as Marx showed in his 'Critique of the Gotha Programme' – is characterised by the fact that social relations will, for a time, necessarily continue to be constrained by the 'narrow horizon of bourgeois right'.[26] It is interesting to analyse what constitutes, in Marx's view, this narrow horizon of bourgeois right (Recht). Marx presupposes a social order in which the means of production are socially owned and in which the producers do not exchange their products. Consequently, he assumes a higher stage of development than the 'new economic policy' which we are presently experiencing.[27] He sees the

---

[26] Marx, 'Marginal Notes to the Programme of the German Workers' Party', in Marx and Engels, Selected Works, vol. III, 1970, p. 19.
[27] 'New Economic Policy': The N.E.P. was introduced in 1921 and established a mixed economy with a free market. It was at an end by 1928 with the establishment of the first Five Year Plan. [Ed.]

market as having been already replaced by an organised framework, such that in no way

> does the labour employed on the products appear here *as the value* of these products, as a material quality possessed by them, since now, in contrast to capitalist society, individual labour no longer exists in an indirect fashion but directly as a component part of the total labour.[28]

Nevertheless, Marx says, even when the market and market exchange have been completely abolished, the new communist society will of necessity be

> in every respect, economically, morally and intellectually, still stamped with the birth marks of the old society from whose womb it emerges.[29]

This becomes evident in the principle of distribution as well, according to which

> the individual producer receives back from society – after the deductions have been made – exactly what he gives to it.[30]

Marx stresses that in spite of the radical transformation of form and content

> the same principle prevails as that which regulates the exchange of commodities, as far as this is exchange of equal values . . . a given amount of labour in one form is exchanged for an equal amount of labour in another form.[31]

To the extent that the relation of the individual producer to society still retains the form of the exchange of equivalents, it also retains the form of law (*Recht*), for 'right' (*Recht*) by its very nature can consist only in the application of an equal standard.'[32] However, as this makes no allowance for the natural inequality of individual talent, it is 'a right of inequality, in its content, like every right'.[33] Marx does not mention that there must be a state authority which guarantees the enforcement of these norms of 'unequal' right by its coercion, thus retaining

---

28 Marx, 'Marginal Notes', op. cit., p. 17.
29 Ibid.
30 Ibid.
31 Ibid., p. 18.
32 Ibid.
33 Ibid.

its 'bourgeois limit', but that goes without saying. Lenin concludes:

> Of course, bourgeois right in regard to the distribution of *consumer* goods inevitably presupposes the existence of the *bourgeois state*, for right is nothing without an apparatus capable of enforcing the observance of the standards of right.
>
> It follows that under communism there remains for a time not only bourgeois right, but even the bourgeois state, without the bourgeoisie![34]

Once the form of equivalent exchange is given, then the form of law – the form of public, or state authority – is also given, and consequently this form persists even after the class structure has ceased to exist. The withering away of law, and with it, of the state, ensues, in Marx's view, only after 'labour has become not only a means of life but life's prime want',[35] when the productive forces grow together with the all-round development of the individual, when everyone works spontaneously according to their abilities, or – as Lenin puts it – when one will no longer be forced to 'calculate with the heartlessness of a Shylock whether one has not worked half an hour more than somebody else',[36] in a word, when the form of the equivalent relation has been finally dispensed with.

Thus Marx conceives of the transition to developed communism not as a transition to new forms of law, but as a withering away of the legal form as such, as a liberation from that heritage of the bourgeois epoch which is fated to outlive the bourgeoisie itself.

At the same time, Marx reveals that the fundamental condition of existence of the legal form is rooted in the very economic organisation of society. In other words, the existence of the legal form is contingent upon the integration of the different products of labour according to the principle of equivalent exchange. In so doing, he exposes the deep interconnection between the legal form and the commodity form. Any society which is constrained, by the level of development of its productive forces, to retain an equivalent relation between expendi-

---

[34] Vladimir Il'ich Lenin, 'The State and Revolution', Chapter 5, in V. I. Lenin, *Selected Works*, Moscow: Progress Publishers, 1968, p. 335.
[35] Marx, 'Marginal Notes', op. cit., p. 19.
[36] Lenin, 'The State and Revolution', op. cit., p. 333.

ture and compensation of labour, in a form which even remotely suggests the exchange of commodity values, will be compelled to retain the legal form as well. Only by starting from this fundamental aspect can one understand why a whole series of other social relations assume legal form. To draw the inference from this, however, that there must always be laws and courts, since not even with the greatest possible economic provision would there be an end to all offences against the person, would simply mean taking secondary, minor aspects for the main, fundamental ones. Even progressive bourgeois criminology has become convinced that the prevention of crime may properly be viewed as a medical-educational problem. To solve this problem, jurists, with their 'evidence', their codes, their concepts of 'guilt', and of 'full or diminished responsibility', or their fine distinctions between complicity, aiding and abetting, instigation and so on, are entirely superfluous. And the only reason this theoretical conviction has not yet led to the abolition of penal codes and criminal courts is, of course, that the overthrow of the legal form is dependent, not only on transcending the framework of bourgeois society, but also on a radical emancipation from all its remnants.

The critique of bourgeois jurisprudence from the standpoint of scientific socialism must follow the example of Marx's critique of bourgeois political economy. For that purpose, this critique must, above all, venture into enemy territory. It should not throw aside the generalisations and abstractions elaborated by bourgeois jurists, whose starting point was the needs of their class and of their times. Rather, by analysing these abstract categories, it should demonstrate their true significance and lay bare the historically limited nature of the legal form.

Every ideology dies together with the social relations which produced it. This final disappearance is, however, preceded by a moment when the ideology, suffering the blows of the critique directed at it, loses its ability to veil and conceal the social relations from which it emanated. The exposure of the roots of an ideology is a sure sign of its imminent end. For, as Lassalle said: 'The dawning of a new age always consists only in the knowledge attained about the true nature of the preceding reality.'[37]

[37] Ferdinand Johann Gottlieb Lassalle, *Das System der erworbenen Rechte: Eine Versöhnung des positiven Rechts und der Rechtsphilosophie*, Leipzig, 1861.

1. The Methods of Constructing the Concrete in the Abstract Sciences

In the study of its subject, every science which makes general-isations is dealing with one and the same concrete and total reality. One and the same observation, such as the observation of a heavenly body moving through the meridian, can give rise to psychological as well as to astronomical inferences. One and the same fact, such as the leasing of land, can form the sub-ject of juridical as well as economic investigations. Thus the various sciences differ mainly in their approach to reality and in their methods. Each science has its own particular design, in terms of which it attempts to reproduce reality. In so doing, each science constructs concrete reality in all its wealth of forms, internal relations and dependencies as the result of a combination of the most simple abstractions. Psychology tries to break down consciousness into its most simple elements. Chemistry applies the same approach to matter. Wherever we are unable in practice to break reality down into its most simple elements, abstraction comes to our aid. It has a particularly large role to play in the social sciences. The maturity of a social science is determined by the degree of perfection attained in the relevant abstraction. Marx illustrates this in an impressive manner with reference to political economy.

He maintains that it would seem quite natural to begin one's analysis with the concrete totality, with the population living and producing in a particular geographical environment; but that population is an empty abstraction if one leaves out the classes of which it is composed. These classes in turn mean nothing without the conditions of their existence such as wages, profit, rent, and so forth. The analysis of the latter presupposes the elementary categories of 'price', 'value', and 'commodity'.

Taking these simplest determinants as his starting point, the economist reproduces the same concrete totality, not any longer, however, as a chaotic, blurred whole, but 'as a rich totality of many determinations and relations'.[1] Marx adds, moreover, that the historical development of economic science has followed precisely the opposite course: the economists of the seventeenth century started from the concrete – nation, state, population – in order to deduce rent, profit, wages, price and value. Yet the fact that it was historically unavoidable to proceed in this manner does not make it methodologically correct.

These observations are directly pertinent to the general theory of law. The concrete totality – society, the population, the state – must in this case, too, be the conclusion and end result of our deliberations, but not their starting point. By moving from the most simple to the more complex, from the process in its purest form to its more concrete manifestations, one is following a course which is methodologically more precise and clearer, and thus more correct, than if one were to feel one's way forward with nothing more in mind than a hazy and unarticulated picture of the concrete whole.

The second methodological observation which it is necessary to make at this point concerns something which is specific to the social sciences, or rather to the concepts used by them. Take, for example, some concept from the natural sciences, such as energy. We can of course establish the exact point in time when it first occurred. Nevertheless, such a date means something only to the history of science and culture. In scientific research itself, the application of this concept is not at all predicated on any time limits. The law of the transformation of energy was in effect before man appeared on earth, and it will continue to take effect after all life on earth is extinct. It exists outside of time; it is an eternal law. One can indeed ask when the law of the transformation of energy was discovered, but it would be futile to raise the question of when the relations which it expresses date from.

[1] Karl Marx, *Grundrisse: Introduction to the Critique of Political Economy*, 1973 ed., p. 100 [Where Pashukanis cites 'wages, profit, rent', Marx in fact puts 'wage labour, capital', and where Pashukanis has 'price, value and commodity', Marx puts 'exchange, division of labour, prices'; that is to say Pashukanis tends in this paragraph to change the categories from production relations to revenues. Ed.]

Turning now to the social sciences, to political economy for instance, and considering one of its fundamental concepts, such as value, it is immediately obvious that this concept not only has an intellectual history, but that, associated with the history of this concept, as part of the history of economic theory, is a real history of value as well, a development in social relations which has gradually turned the concept into historical reality.[2]

We know exactly which material conditions are necessary for this 'hypothetical', 'imaginary' quality of things to gain a 'real' – and indeed decisive – significance as compared with their natural properties, transforming the product of labour from a natural into a social phenomenon. Thus we are familiar with the real historical substratum of those cognitive abstractions we use, and we can, at the same time, satisfy ourselves that the limits within which the application of these abstractions is meaningful are synonymous with, and are determined by, the framework of actual historical development. Another example cited by Marx illustrates this particularly graphically. Labour as the simplest relation of man to nature is present at all stages of development without exception; yet as an economic abstraction it is relatively recent (compare the sequence of schools of thought: the Mercantilists, the Physiocrats, the Classical economists). This development of the concept was paralleled by the actual development of economic relations, a development which pushed aside the diversity of human labour, replacing it with 'labour in general'. Thus the development of the concepts corresponds to the actual dialectic of the historical process.[3]

Let us consider another example, this time not from the field of political economy. Take the state. Here we can observe how, on the one hand, the concept of the state gradually acquires precision and definitiveness and develops the full potential of its determinants, and yet, on the other hand, how

---

[2] It is a mistake, however, to imagine that the value form and the theory of value evolved synchronously. On the contrary: these two processes did not coincide in time at all. More or less developed forms of exchange and the corresponding value forms are to be found in the most distant antiquity, whilst political economy is, as everyone knows, one of the youngest sciences. (Note to the 3rd Russian ed.)

[3] Marx, *Grundrisse*, ed. cit., pp. 104-105.

the state in reality grows out of the gens-community[4] and feudal society, 'abstracts itself' and transforms itself into a 'self-sufficient' force, 'blocking up all the pores of society'.[5]

Hence law in its general definitions, law as a form, does not exist in the heads and the theories of learned jurists. It has a parallel, real history which unfolds not as a set of ideas, but as a specific set of relations which men enter into not by conscious choice, but because the relations of production compel them to do so. Man becomes a legal subject by virtue of the same necessity which transforms the product of nature into a commodity complete with the enigmatic property of value.

To the kind of thinking which does not transcend the framework of the bourgeois conditions of existence, this necessity must appear as none other than natural necessity. That is why all bourgeois theories of law are based, consciously or unconsciously, on the doctrine of natural law. The school of natural law was not only the most marked expression of bourgeois ideology at the time when the bourgeoisie acted as a revolutionary class, formulating its demands openly and consistently; it also provided the model for the deepest and clearest understanding of the legal form. It is not by chance that the period when the doctrine of natural law flourished coincides approximately with the appearance of the great classical bourgeois political economists. Both schools set themselves the task of formulating, in the most general, and consequently the most abstract manner, the fundamental conditions of existence of bourgeois society, which they regarded as the natural conditions of existence for absolutely any society.

Even someone so overzealous in the cause of legal positivism and so opposed to natural law as Bergbohm feels bound to acknowledge the achievements of the natural law school in laying the foundations of the modern bourgeois legal system.

It (natural law, E.P.) threatened serfdom and bondage and pressed for an end to people's enslavement to the land and the soil; it unleashed the productive forces which had been fettered

[4] Cf. Friedrich Engels, 'Origins of the Family, Private Property and the State', in Karl Marx and Friedrich Engels, *Selected Works*, vol. III, 1970, p. 326. [Ed]

[5] This is Marx, 'The Eighteenth Brumaire of Louis Bonaparte', in Marx and Engels, *Selected Works*, vol. I, 1969, p. 477. [Ed.]

by the coercion of an ossified system of guilds and by absurd trade restrictions . . . it achieved freedom of religious persuasion as well as the freedom of scientific teaching . . . it gained the protection of private law for every religion and every nationality . . . it helped to abolish torture and to guide the criminal case into the ordered course of a procedure according to law.[6]

While not entertaining the intention of concerning ourselves in full detail with the sequence of the various schools in the theory of law, nevertheless, we cannot avoid drawing attention to a certain parallel in the development of juridical and economic thought. Thus the historical tendency can be considered in both cases as a manifestation of the feudal-aristocratic and, in part, of the petty bourgeois guild reaction. Further: when, in the second half of the nineteenth century, the revolutionary zeal of the bourgeoisie finally faded out, the purity and precision of the classical doctrines simultaneously lost all attraction for it. Bourgeois society yearns for stabilisation and a strong arm. That explains why it is no longer the analysis of the legal form, but the problem of justifying the binding force of legal regulations which becomes the focal point of interest for juridical theory. The result is a strange mixture of historicism and juridical positivism which is reduced to negating every law except the official law.

The so-called 'renascence of natural law' does not signify that bourgeois legal philosophy is reverting to the revolutionary views of the eighteenth century. In Voltaire's and Beccaria's time every enlightened judge counted it as an achievement if, in the guise of applying the law, he succeeded in substantiating the ideas of the philosophers, ideas which were no less than a revolutionary negation of the feudal social order. In the present day, the prophet of renascent 'natural law', Rudolf Stammler, advances the thesis that 'true law' (*richtiges Recht*) requires first and foremost subjection to positively promulgated law, even if it be 'unjust' (*ungerecht*).

A parallel can be drawn between the psychological school of jurisprudence and the psychological school of political economy. Both are at pains to transfer the object of analysis

[6] Carl Bergbohm, *Jurisprudenz und Rechtsphilosophie: Kritische Abhandlungen*, Leipzig, 1892, vol. I, p. 215.

into the realm of subjective areas of consciousness ('evaluation', 'imperative-attributive emotion'), and fail to see that the ordering of the corresponding abstract categories expresses the logical structure of social relations which are concealed behind individuals and which transcend the bounds of individual consciousness.

Finally, there is no doubt but that the extreme formalism of the normative school (Kelsen) expresses the general decadence of the most recent bourgeois thinking, which spends itself in sterile methodological and formal-logical humbug and parades its own complete dissociation from actual reality. In economic theory, the representatives of the mathematical school would fill the corresponding position.

The legal relation is, to use Marx's expression, an abstract, one-sided relation, which is one-sided not as a result of the intellectual labour of a reflective subject, but as the product of social development.

> In the succession of economic categories, as in any other historical, social science, it must not be forgotten that their subject – here, modern bourgeois society – is always what is given, in the head as well as in reality, and that these categories therefore express the forms of being, the characteristics of existence, and often only individual sides of this specific society, this subject.[7]

What Marx says here about economic categories is directly applicable to juridical categories as well. In their apparent universality, they in fact express a particular aspect of a specific historical subject, bourgeois commodity-producing society.

In conclusion, we find in the same 'Introduction' by Marx from which we have already quoted so freely yet another profound methodological observation. It concerns the possibility of clarifying the significance of earlier structures through the analysis of later and consequently more highly developed ones. If we understand ground-rent, he says, we can also understand tribute, tithes and feudal dues. The more highly developed form renders the prior stages, in which it appears only as an embryo, comprehensible to us. The later evolution simultaneously reveals the intimations implicit in the distant past.

---

[7] Marx, *Grundrisse*, ed. cit., p. 106.

Bourgeois society is the most developed and the most complex historic organisation of production. The categories which express its relations, the comprehension of its structure, thereby also allows insights into the structure and the relations of production of all the vanished social formations out of whose ruins and elements it built itself up, whose partly still unconquered remnants are carried along within it, whose mere nuances have developed explicit significance within it, etc.[8]

Applying these methodological considerations to the theory of law, we must start with an analysis of the legal form in its most abstract and pure shape and then work towards the historically concrete by making things more complex. In the process, we must not lose sight of the fact that the dialectical development of the concepts parallels the dialectic of the historical process itself. Historical development is accompanied not only by a transformation of the content of legal norms and legal institutions, but also by development in the legal form as such. Having emerged at a particular stage of culture, this legal form persists for a long time in an embryonic state, with minimal internal differentiation, and with no clear demarcation from neighbouring spheres (mores, religion). Only after a period of gradual development does it reach its full flowering, its maximum differentiation and definition. This highest stage of development corresponds to quite specific economic and social relations. It is characterised simultaneously by the emergence of a set of general concepts which comprise a theoretical reflection of the legal system as a perfected whole.

Corresponding to these two cycles of cultural development, there are two epochs when the general concepts of law reached their highest point of development: Rome, with its system of private law, and the seventeenth and eighteenth centuries in Europe, during which time philosophical thought discovered the universal significance of the legal form as a possibility which bourgeois society was destined to embody.

It follows that we can reach clear and exhaustive definitions only by basing our analysis on the fully developed legal form, which recognises itself in embryo in preceding legal forms.

Only then shall we comprehend law not as an appendage of

[8] Ibid, p. 105.

human society in the abstract, but as an historical category corresponding to a particular social environment based on the conflict of private interests.

2. Ideology and Law

The question of the ideological nature of law played an important part in a polemic between Stuchka and Reisner.[1] Professor Reisner tried to establish that Marx and Engels themselves considered law as one of the 'ideological forms', and that many other Marxist theoreticians held the same view, supporting his argument with an impressive number of quotes. One cannot quibble with these references and quotations, just as one cannot question the fact that people experience law at a psychological level, especially when it figures as general norms or regulations of principle. However, it is not a matter of affirming or denying the existence of the ideology (or psychology) of law, but rather of demonstrating that the categories of law have absolutely no significance other than an ideological one. Only if this were established could we accept as incontrovertible Professor Reisner's conclusion, namely that 'a Marxist can study law only as a sub-category of the species Ideology'. The little word 'only' is the crux of the matter. We shall illustrate this with an example from political economy. The categories commodity, value and exchange value are indubitably ideological constructs, distorted, mystified mental images (as Marx puts it), by means of which the society based on the exchange of commodities conceives of the labour relation between individual producers. The ideological nature of these forms is proven by the fact that, no sooner do we come to other forms of production than the categories of the commodity, value and so on cease to have any validity whatever. Consequently we are justified in speaking of a commodity-orientated ideology, or, as Marx called it, 'commodity fetishism', and in classing this

[1] Cf. *Vestnik Sotsialisticheskoy Akademii*, no. 1.

phenomenon as a psychological one.[2] Nevertheless, this in no way implies that the categories of political economy have an *exclusively* psychological significance, or that they relate *solely* to experiences, representations, and other subjective processes. We are well aware that the category 'commodity' for instance, its blatantly ideological character notwithstanding, reflects an objective social relation. We also know that the various stages of development of this relation, its greater or lesser degree of universality, are material facts which must be taken into account as such and not merely as ideological-psychological processes. It follows then that the general concepts of political economy are not merely ideological factors; rather they are abstractions of a kind which enables objective economic reality to be scientifically, that is theoretically, constructed. To quote Marx:

> they are forms of thought which are socially valid, and there-fore objective, for the relations of production belonging to this historically determined mode of social production, i.e. commodity production.[3]

What we need to establish, therefore, is not whether general juridical concepts can be incorporated into ideological processes and ideological systems – there is no argument about this – but whether or not social reality, which is to a certain extent mystified and veiled, can be discovered by means of these concepts. In other words: we must be clear about whether or not the categories of law are objective forms of thought (objective for the historically given society) corresponding to the objective social relations. Hence we formulate the question as follows: *Can law be conceived of as a social relation in the same sense in which Marx called capital a social relation?*

Putting the question in this way immediately makes any reference to the ideological nature of law unnecessary and transposes the whole analysis onto a different plane.

Having established the ideological nature of particular concepts in no way exempts us from the obligation of seeking their objective reality, in other words the reality which exists

---

[2] Cf. Karl Marx, *Capital*, vol. I, 1976 ed., pp. 164-169. [Ed.]
[3] Ibid., p. 169.

in the outside world, that is, external, and not merely subjective reality. Otherwise any distinction between life after death, which does exist in some people's minds, and, let us say, the state, would be obliterated. Yet this is exactly where Reisner's approach leads. He bases himself on the well-known Engels quotation about the state as 'the first ideological power over man',[4] and has no hesitation in equating the state with its ideology.

> The psychological nature of manifestation of power is so obvious, and even state power, which *only exists in the human psyche* (my underlining, E.P.), is so lacking in material characteristics, that one would think no-one could see state power as anything but an Idea which only materialises in real terms to the extent that people make it the principle governing their behaviour.[5]

The Treasury, the military, the administration, all these supposedly lack material characteristics, all this only exists 'in the human psyche'. Yet what is supposed to become of that mass of the population – 'colossal' in Reisner's own words – which lives 'beyond consciousness of the state'? Obviously this mass will have to be excluded; indeed it seems to have no bearing whatever on the 'real' existence of the state.

How, then, do matters stand with the state as an economic entity? Or is the tariff barrier an ideological and psychological process as well? One could raise a great many questions of this kind, yet they would all come to the same thing. The state is not merely an ideological form, but is at the same time a form of social being. The ideological nature of the concept does not obliterate the reality and the material nature of the relations which it expresses.

It is understandable that Kelsen, the consistent Neo-Kantian, should maintain the normative, the purely speculative, objectivity of the state and that he should jettison, not only concrete-material factors, but even the actual human psyche as

---

[4] Friedrich Engels, 'Ludwig Feuerbach and the End of Classical German Philosophy', in Marx and Engels, *Selected Works*, vol. III, 1970, p. 371. [Ed.]
[5] Mikhail Andreyevich Reisner, *The State (Gosudarstvo)*, 2nd ed., Moscow, 1918, Part I, p. 35.

well. But there is no way we can comprehend a Marxist (materialist) theory which tries to operate on the basis of subjective experiences alone. Moreover, Reisner, whose latest works show him to be an advocate of Petrazhitsky's psychological theory (a theory which comple⁺ely 'carves up' the state into a set of imperative-attributive emotions), would not be adverse to linking this viewpoint with Kelsen's Neo-Kantian formal-logical conception.[6] At all events, such an attempt does credit to the versatility of our author, even if it be undertaken at the expense of methodological consistency and clarity. One thing or the other: either the state is an ideological process (Petrazhitsky), or, (as Kelsen maintains), it is a regulative Idea having absolutely nothing to do with any processes which occur in time and are subject to the law of causality. In trying to combine these two positions, Reisner lapses into a completely undialectical contradiction.

The formal completeness of the concepts 'state territory', 'population', 'state authority' reflects, not only a particular ideology, but also the objective fact of the formation of a tightly-centred real sphere of dominance and thus reflects, above all, the creation of an actual administrative, fiscal and military organisation with the corresponding material and human apparatus. The state is nothing without means of communication, without the possibility of transmitting orders and decrees, mobilising the armed forces, and so on. Does Reisner think that Roman army roads, or modern means of communication, can be numbered among the phenomena of the human psyche? Or does he think that these material elements simply need not be taken into account as factors affecting the formation of the state? If so, we should indeed be equating the reality of the state with the reality of 'literature, philosophy and other products of the intellect'.[7] What a pity that the practice of the political power struggle radically contradicts this psychological view of the state, confronting us at every step with objective material factors.

This leads us to the observation that the inevitable outcome

[6] Reisner, 'Social Psychology and Freudian Theory' ('Sotsial'naya psikologiya i uchenie freyda'), in *Press and Revolution, (Pechat i revolyutsiya)*, Moscow, 1925, vol. II.
[7] Reisner, *The State*, ed. cit., p. 48.

of the psychological standpoint adopted by Reisner is hopeless subjectivism which leads nowhere.

State power as the creature of as many psychologies as there are individuals, state power which manifests as many different models as there are diverse group- and class-based environments, will naturally assume a completely different shape in the consciousness and behaviour of a minister and that of a peasant as yet unfamiliar with the idea of the state; in the psyche of a statesman and that of a convinced anarchist. In a word, people's conception of the state varies in accordance with their social status, their trade and their level of education.[8]

It is abundantly clear from this that to remain on the psychological plane is to forfeit all grounds for speaking of the state as an objective entity. Only by regarding the state as a real organisation of class rule (taking all its aspects into account – not the psychological alone, but also, and above all, the material aspects) does one acquire a firm basis for studying the state as it really is, rather than merely the countless and varied subjective forms through which it is reflected and experienced.[9]

[8] Ibid., p. 35.
[9] Professor Reisner seeks substantiation of his view (cf. his 'Social Psychology and Freudian Theory') in a letter from Friedrich Engels to Conrad Schmidt, in which Engels investigates the problem of the connection between phenomenon and concept. Taking the feudal social order as his example, Engels points out that the unity of phenomenon and concept presents itself as an intrinsically endless process. 'Did feudalism ever correspond to its concept? . . . Was this order a fiction because it existed in full classical form only in Palestine, and even there was short-lived and mostly only on paper at that?' [This translation is from the German. The English edition is: Engels to C. Schmidt, March 12, 1895, in Marx, Engels, *Selected Correspondence*, 2nd ed., 1965, p. 484.] Nevertheless, it in no way follows from Engels' remarks that equating phenomenon and concept, as Reisner does, is correct. For Engels, the concept of feudalism and the feudal social order were in no way identical. On the contrary, Engels proves that feudalism never actually corresponded to its concept, yet for all that it did not cease to be feudalism. The concept of feudalism is an abstraction based on actual tendencies of that social order which we call feudal. In historical reality, these tendencies mingle and intersect with innumerable other tendencies and therefore cannot be observed in their logically pure form, but only in greater or lesser approximation to it. Engels alludes to this, in saying that the unity of phenomenon and concept is basically an infinite process.

Given that these abstract definitions of the legal form not only imply certain psychological or ideological processes, but are concepts which express objective social relations, in what sense can it be said that law regulates social relations? Surely we do not wish to imply that social relations are self-regulating? For if we say that this or that social relation takes on legal form, this is not meant to be a mere tautology: law assumes legal form.[10]

At first sight this objection seems very convincing and appears to leave no alternative but to acknowledge that law is ideology. Nonetheless, we shall endeavour to find a way out of these difficulties. So as to make this task easier, we shall take refuge yet again in a comparison. As we know, Marxist political economy holds that capital is a social relation. As Marx says, it cannot be discovered by means of a microscope, nor is it exhaustively dealt with in experiences, ideologies and other subjective processes occurring in the human psyche. It is an objective social relation. Further, when we observe, let us say in petty commodity production, a gradual transition from labour for a customer to labour for an entrepreneur, we conclude that the corresponding relations have assumed capitalist form. Does this mean that we have lapsed into a tautology? Not at all; in saying this we have simply stated that the social relation known as capital has tinted, or transferred its own form to a different social relation. We are able to establish this by considering all the processes involved purely from the objective side, as material processes, eliminating completely the psychology or ideology of the participants. Is there any reason why the situation should not be just the same with regard to law? As it is a social relation in itself, it is capable of colouring other social relations to a greater or lesser degree, or of transmitting its form to them. However, we can never gain access to the problem from this angle if we allow ourselves to be guided by an unclear notion of law as 'form as such', just as vulgar political economy was not able to grasp the nature of capitalist relations, because it started from the concept of capital as 'accumulated labour in general'.

Consequently we escape the apparent contradiction if we

10 Cf. Reisner's review of Stuchka's book in *Vestnik Sotsialisticheskoy Akademii*, no. 1, p. 176.

succeed in establishing, by analysing its fundamental definitions, that law represents the mystified form of a specific social relation. It would not then be far-fetched to assert that in certain cases this relation transmits its own form to some other social relation, or even to the totality of social relations.

The same applies to the second apparent tautology, according to which law regulates social relations. If one strips this formula of a certain anthropomorphism attached to it, it reduces itself to the following: under certain conditions the *regulation* of social relations assumes a *legal character*. This is undoubtedly a more correct way of putting it and, most importantly, it is historically more accurate. There is no denying that there is a collective life among animals too which is also regulated in one way or another. But it would not occur to us to assert that the relations of bees or ants are regulated *by law*. Turning to primitive peoples, we do see the seeds of law in them, but the greater part of their relations are regulated extra-legally, by religious observances for instance. Finally, even in bourgeois society there are things like the organisation of the postal and rail services, of the military, and so on, which cannot be related in their entirety to the sphere of *legal* regulation unless one views them very superficially and allows oneself to be confused by the outward form of laws, statutes and decrees. Train timetables regulate rail traffic in quite a different sense than, let us say, the law concerning the liability of the railways regulates its relations with consigners of freight. The first type of regulation is predominantly technical, the second primarily legal. The same distinction exists between a plan for mobilisation and the law covering universal conscription, between a brief for investigating crime and criminal proceedings.

We will come back to the difference between technical and legal norms in what follows. For the moment we merely make the observation that the regulation of social relations can assume legal character to a greater or lesser extent, can allow itself to be more or less coloured by the fundamental relation specific to law.

Only when observed in a superficial or merely formal manner does the regulation or standardisation of social relations appear fact there are very striking differences in this respect between as a fundamentally homogeneous and purely legal process. In the various spheres of human relations. Gumplowicz first drew a sharp dividing line between private law and state norms, but in so doing he tried to admit of only the former sphere as the undisputed domain of jurisprudence.[11] In reality, the hardest core of legal haziness (if one may be permitted to use such an expression) is to be found precisely in the sphere of civil law. It is here above all that the legal subject, the *'persona'*, finds entirely adequate embodiment in the real person of the subject operating egoistically, the owner, the bearer of private interests. Juridical thought moves most freely and confidently of all in the realm of private law; its constructs assume perfect and well-ordered forms. Here the classical shades of Aulus Aegerius and Numerius Negidius, those Roman protagonists of procedural questions who are the inspiration of the jurist hover over him constantly. It is above all in private law that the *a priori* principles and premises of juridical thought become clothed in the flesh and blood of two litigating parties who, *vindicta* in hand,[12] claim 'their right'. Here the role of the jurist as theoretician merges with his practical social function. The doctrine of private law is no more than an endless chain of deliberations for and against hypothetical claims and potential suits. Behind every paragraph of the systematic thread of the argument stands the invisible, abstract client, ready to utilise the theses in question as legal advice. The learned disputes between academic jurists about the significance of error, or about the allocation of the onus of proof are no different from similar disputes before the courts. The distinction here is no greater than that which used to exist between knightly tournaments and feudal battles. The former took place, as we know, with great bitterness, at times claiming no lesser expenditure of energy, nor fewer victims, than real affrays. Only when the individualistic economic system has been superseded by planned social production and distribution will this unproductive expenditure of man's intellectual energies cease.[13]

[11] Cf. Ludwik Gumplowicz, *Rechtsstaat und Sozialismus*, Innsbruck, 1881.
[12] 'Vindicta': Originally the liberating rod with which a slave was touched during the ceremony of manumission, the word generally came to signify a means of asserting or defending – a protection or defence (its most common usage was: 'vindication' of liberty). [Ed.]
[13] A short piece by T. Yablochkov, 'Adjournment and the Burden of

A basic prerequisite for legal regulation is therefore the conflict of private interests. This is both the logical premise of the legal form and the actual origin of the development of the legal superstructure. Human conduct can be regulated by the most complex regulations, but the juridical factor in this regulation arises at the point when differentiation and opposition of interests begin. Gumplowicz states that 'controversy is the fundamental element of everything juridical'. In contrast to this, the prerequisite for technical regulation is unity of purpose. For this reason the legal norms governing the railways' liability are predicated on private claims, private, differentiated interests, while the technical norms of rail traffic presuppose the common aim of, say, maximum efficiency of the enterprise. To take another example: healing a sick person presupposes a set of rules, for the patient as well as for the medical personnel. In so far as these rules have been prescribed for the express purpose of rehabilitating the sick person, they are technical in nature. The enforcement of these rules can be associated with some degree of constraint on the sick person. So long as this constraint is viewed from the standpoint of a goal which is the same for the person exercising the coercion as it is for the person coerced, it is a technically expedient act and no more. The content of the regulations is specified within these limits by medical science and undergoes change as medical science progresses. The lawyer has no place here. His role begins at the point where we are forced to leave

Proof' ('Suspensivnoe Uslovie I. Bremya Dokazyvaniya', in *Yuridichesky Vestnik*, 1916, no. 15, p. 55), testifies to what significant levels this wastage of human acumen has reached. In it, he portrays the history and literature on the single legal problem concerning the distribution of proof between the parties, by means of the accused appealing for adjournment. The author cites no less than fifty learned men who have written on the subject in a literature going right back to the post-glossators. Yablochkov informs us that two 'theories', set up to decide the question, split the entire academic legal world into two approximately equal camps. He is delighted by the exhausting mass of arguments marshalled by both sides as long as a hundred years ago (which has obviously not prevented later researchers into this problem from reproducing the same arguments in different tones), and pays tribute to the 'far-reaching analysis' and the 'acumen of the methodological procedures' of the learned disputants. He informs us that the dispute so inflamed the passions that, in the heat of battle, the adversaries accused one another of slander, spreading false rumours, and mutually accused their respective theories of being dishonest and unethical.

this realm of unity of purpose and to take up another stand-point, that of mutually opposed separate subjects, each of whom represents his own private interests. Doctor and patient are thereby transformed into subjects with rights and duties, and the regulations which govern them are transformed into legal norms. Simultaneously with this, coercion is no longer considered under the rubric of expediency, but from the point of view of formal, that is of legal, admissibility.

It is not hard to see that the possibility of taking up a legal standpoint is linked with the fact that, under commodity pro-duction, the most diverse relations approximate the prototype commercial relation and hence assume legal form. Similarly, it is a matter of course for bourgeois jurists to infer the univer-sality of the legal form, either from eternal and absolute charac-teristics of human nature, or from the fact that official decrees can be arbitrarily applied to any subject. It is hardly necessary to give detailed evidence of the latter. In the Civil Code of the pre-revolutionary Russian Empire there was, after all, an article imposing on the husband the duty 'to love his wife as he loves his own body'. Yet even the most presumptuous jurist would hardly have attempted to construe a legal relation allowing for the possibility of suits on this basis.

On the contrary, no matter how ingeniously devised and un-real any one of the juridical constructs may appear, it is on firm ground so long as it remains within the bounds of private law, and of property law in particular. It would otherwise be impossible to grasp the fact that the fundamental trains of thought of Roman jurists have retained their significance up to the present day and have remained the *ratio scripta* of every commodity-producing society.

In the above, we have to a certain extent anticipated the answer to the question raised at the outset as to whether a social relation *sui generis* is to be found whose inevitable re-flex is the legal form. In what follows, we shall try to establish that this relation is the interrelationship of the owners of com-modities.[14] The usual analysis to be found in absolutely any

---

14 Cf. Vladimir Viktorovich Adoratsky, *On the State* (*O gosudarstve*), Moscow, 1923, p. 41: 'The colossal influence of legal ideology on the whole thought process of orthodox members of bourgeois society is based on the enormous role legal ideology plays in the life of this

philosophy of law construes the legal relation as a relation par excellence, as a relation of human wills in general. This view starts out from the 'end results of the developmental process', from 'current modes of thought', without accounting for their historical origin: whereas in reality, as commodity-production develops, the natural prerequisites for the act of exchange become the natural premises, or natural forms, of all social intercourse. Acts of trade are seen by the philosophers, in contrast to this, merely as particular examples of the general form which in their minds has assumed an eternal quality.[15]

Comrade Stuchka has quite rightly posed the problem of law as a problem of social relations. Yet instead of setting out in search of the objective social reality specific to this relation, he turns back to the usual formal definition, albeit qualified by class determinants. In Stuchka's general formulation, law no longer figures as a *specific* social relation, but *as the sum of relations in general, as a system of relations corresponding to the interests of the ruling class and to the safeguarding of those interests by organised violence.* It follows therefore that, within this class framework, law as a relation is indistinguishable from social relations in general, and Stuchka is no longer in a position to parry Professor Reisner's malicious question as to how social relations were transformed into legal institutions or, for that matter, how law transformed itself into itself.[16]

---

society. The exchange relation takes place in the form of legal transactions of buying and selling, lending and pawning, rent and so forth'. And: 'The person living in bourgeois society is constantly considered as the subject of rights and obligations; daily he commits an infinite number of legal acts, which incur the most diverse legal consequences. Therefore no society needs the idea of law so badly, just for practical daily usage, as does bourgeois society, no other society submits this category to such detailed elaboration, nor transforms it into so indispensable a means of everyday transactions.'

[15] Marx, *Capital*, vol. I, op. cit., p. 168.

[16] Stuchka thinks that gave an explanation of this point as early as a year before the publication of my book (Cf. Petr Ivanovich Stuchka, *The Revolutionary Part Played by Law and the State: A General Doctrine of Law, Revolyutsionnaya rol' prava i gosudarstva*, 3rd ed., Moscow, 1924, p. 112; [an English translation of this work is included in *Soviet Legal Philosophy*, 1951, pp. 17-69. Transl] Law as a specific set of social relations is distinguished, according to him, by the fact that it is maintained by the organised violence of one class, that is by state power. I was of course familiar with this view, but I still think, even after a second elucidation, that in a set of relations cor-

Perhaps because it emanated from the womb of the People's Commissariat for Justice, Stuchka's definition is adapted to the needs of the practising lawyer. It illustrates the empirical limit which history imposes on juridical logic, but it does not expose the deep roots of this logic itself. His definition uncovers the class content concealed within legal forms, but does not explain why this content assumes that particular form.

For bourgeois philosophy, which regards the legal relation as the eternal, natural form of every human relation, this question never even arises. For Marxist theory, which strives to penetrate the secrets of social forms and to reduce all social relations to man himself, this task must take priority.

---

responding to the interests of the ruling class and upheld by the organised force of that class, there are factors which can and should be distinguished as providing material pre-eminently suited to the development of the legal form.

3. Norm and Relation

In as much as the wealth of capitalist society appears as 'an immense collection of commodities',[1] so this society itself appears as an endless chain of legal relations.

Commodity exchange presupposes an atomised economy. The link between isolated private economic units is maintained in each case by successfully concluded business deals. The legal relation between subjects is simply the reverse side of the relation between products of labour which have become commodities. This does not prevent certain jurists, Petrazhitsky for one, from inverting things and believing, not that the commodity form produces the legal form, but that, on the contrary, the economic phenomena studied by political economy

> represent people's individual and mass behaviour, which is conditioned by a typical motivation emanating from the institutions of civil law (private property, law of obligations and contract, family law and law of inheritance). [2]

The legal relation is the cell-form of the legal fabric; only there does law accomplish its real movement. Compared to this, law as the aggregate of norms is merely a lifeless abstraction.

It follows quite logically that the normative school, headed by Kelsen, should totally deny the relation between subjects, refusing to study law from this point of view, and preferring to concentrate their undivided attention on the formal relevance of the norms.

[1] Karl Marx, *Capital*, vol. I, 1976 ed., p. 125 [Ed.]
[2] Lev Yosifovich Petrazhitsky, *Introduction to the Study of Law and Morality* (*Vvedenie v izuchenie prava i nravstvennosti*), vol. I, p. 77; [abridged form translated by H. W. Bass as: *Law and Morality*, with an introduction by N. S. Timasheff, Cambridge, Mass.: Harvard University Press, 1955 (20th Century Legal Philosophy Series, vol. 7). Transl.]

The legal relation is a relation for the purpose of the legal system or, more correctly: it is a relation within the legal system, not a relation between legal subjects separate from the system of law.[3]

Nonetheless, the current view holds that the legal relation is not only logically, but also in reality, based on the norm, or objective law. According to this notion, the legal relation is generated by the objective norm:

> It is not because creditors generally demand repayment of a debt that the right to make such a demand exists, but, on the contrary, the creditors make this claim because the norm exists; the law is not defined by abstraction from observed cases, but derives from a rule posited by someone.[4]

The expression 'the norm generates the legal relation' can be understood in two senses: real and logical.

Let us consider the former. It must be noted above all – and the jurists themselves have tried often enough to assure one another of this – that the totality of written or unwritten norms actually belongs rather to the sphere of literary creation.[5] This totality of norms only attains real meaning thanks to the relations which are thought to arise in conformity with these regulations and do indeed arise in this way. Even the most consistent follower of the normative method, Hans Kelsen, had to admit that some part of real life, that is to say of people's actual behaviour, must in some way be injected into the ideal normative system.[6] Indeed, only a person ripe for the madhouse would today regard the laws of tsarist Russia as valid

[3] Hans Kelsen, *Das Problem der Souveranität und die Theorie des Völkerrechts: Beitrag zu einer reinen Rechtslehre*, Tübingen, 1920, p. 125.
[4] Shershenevich, *General Theory of Law* (*Obshchaya Teoriya Prava*), 1910, p. 74.
[5] Cf. in Hold von Ferneck: 'We must take account of the fact that laws only generate "law" insofar as they are realised and emancipate themselves as norms from their "paper existence" in order to prove themselves as a force in human life' (Alexander Hold von Ferneck, *Die Rechtswidrigkeit: Eine Untersuchung zu den allgemeinen Lehren des Strafrechtes*, Jena, 1903, p. 11).
[6] Cf. Kelsen, *Der soziologische und der juristische Staatsbegriff: Kritische Untersuchung des Verhältnisses von Staat und Recht*, Tübingen, 1922, p. 96.

law. The formal-juridical method, which is concerned only with norms, only with that 'which is according to law', can maintain its independence only within very narrow limits and even then only so long as the tension between fact and norm does not exceed a certain maximum level. In material reality, the relation has primacy over the norm. If no debtor repaid his debts, the relevant regulation would have to be considered as non-existent in real terms. Should we wish to assert the existence of these regulations despite this, we would be forced to fetishise the norm in some way or other. A great many theories of law are concerned with just such festishisation, which they base on very subtle methodical considerations.

Law as an objective social phenomenon cannot be exhaustively defined by the norm or regulation – whether it be written or unwritten. Turning now to the logical content of the norm as such, it is either derived directly from relations already in existence or, if it is decreed as state law, it represents a mere symptom, from which one can infer with some probability the emergence of corresponding relations in the near future. Nevertheless, to assert the objective existence of law, it is not enough to know its normative content, rather one must know too whether this normative content materialises in life, that is in social relations. The usual source of errors in this case is the legal dogmatist's way of thinking – for him, the specific significance of the concept of the valid norm does not coincide with that which the sociologist or the historian understand by the objective existence of law. When the legal dogmatist has to decide whether or not a particular legal form is valid, most often he makes no atempt at all to ascertain whether a certain objective social phenomenon is present or absent, but only whether or not there is a logical connection between the given normative proposition and the more general normative premise.[7]

Thus for the legal dogmatist, within the narrow bounds of his purely technical task there really is nothing beyond the

[7] Incidentally, the Russian language uses terms etymologically derived from the same root in order to designate 'effective' law and 'valid' law. In German, the logical distinction is made easier through the use of two entirely different verbs : *wirken* (in the sense of to have an effect) and *gelten* (in the sense of to be valid, that is to say, to be linked to a more general normative premise). [Ed.]

norms; he can therefore equate norm with law with the greatest equanimity. With regard to prescriptive right he has to turn to reality, whether he likes it or not. While the law of the state may be the highest normative prerequisite or, to use the technical term, the source of law for the jurist, the legal dogmatist's views on 'valid' law are not in the least binding for the historian who wishes to study law as it actually exists. Scientific, that is, theoretical study can reckon only with facts. If certain relations have actually come into being, this signifies that a corresponding law has arisen; however, if a law or decree has merely been promulgated without any corresponding relation having arisen in practice, then an attempt to create a law has indeed been made, but without success. This position is in no way synonymous with the negation of class will as a factor in development, or with renouncing intervention in the process of social development, with 'economism', fatalism or any other such dreadful things. Revolutionary political action can achieve a great deal; it can realise tomorrow that which does not yet exist today, but it cannot lend existence after the fact to something which did not actually exist in the past. Alternatively, if we assert that the intention to construct a building, and even the plan for it, do not add up to the building itself, it does not follow from this at all that its construction necessitates neither intention nor plan. Yet if the matter has only got as far as the plan and no further, we cannot assert that the building has been erected.

Moreover, one can modify somewhat the thesis (that norm and law can be equated) and emphasise, not the norm as such, but rather the objective regulative forces operating in society or, as the jurists put it, the objective legal system![8]

Yet even this modified formulation of the thesis can be subjected to further criticism. If by socially regulative forces one

[8] We must point out here that a socially regulative activity can also do without norms laid down *a priori*. So-called judicially created law is evidence of this. It has great significance, especially in those epochs which did not know the centralised promulgation of laws. Thus, for instance, the concept of a final, externally given norm was completely unknown to the ancient Germanic courts. To the jurors, all collections of regulations were not binding law, but expedients, on the basis of which they formed their own opinion. (Johann August Roderich von Stintzing, *Geschichte der deutschen Rechtswissenschaft*, 1880, vol. I, p. 39.)

means merely the same relations in their regularity and permanence, one has a simple tautology. But if under that heading one means a particular, consciously organised system which guarantees and safeguards these relations, then the fallacy becomes absolutely clear. Of course one cannot assert that the relation between creditor and debtor is *generated* by the system of compulsory debt collection operating in the state in question. The objective existence of this system certainly *guarantees* and *safeguards* the relation, but it in no way creates it. That this is no mere scholastic semantic quibble can best be shown by the fact that one can conceive of very different degrees of perfection in the functioning of this external coercive social regulation and consequently of the most varying degrees of guarantee of certain regulations (and can substantiate this with historical examples), without these relations themselves suffering the smallest variation in their substance. We can conceive of a borderline case in which, apart from the two parties relating to one another, no other third force can determine a norm and guarantee its observance: for example, any contract of the Varangians with the Greeks. Even in this case the relationship still remains in existence.[9] Nevertheless, one need only imagine the disappearance of one of the parties, one of the subjects as the representative of an autonomous separate interest, for the possibility of the relation itself to vanish too.

It is possible to counter our view by saying that if one abstracts from the objective norm, the concepts of the legal relation and the legal subject themselves have no foundation and cannot be defined at all. This objection expresses the eminently practical empirical spirit of modern jurisprudence, which has one firm conviction: namely that every trial would be lost if the plaintiff were not able to base his case on the

---

[9] The entire feudal legal system rested on such contractual relations, guaranteed by no 'third force'. In just the same way, modern international law recognises no coercion organised from without. Such non-guaranteed legal relations are unfortunately not known for their stability, but this is not yet grounds for denying their existence. There is no absolutely permanent law; on the other hand, the stability of private law relations in the modern, 'well-ordered', bourgeois state does not rest on the police and the courts alone. Debts are paid not only because they 'will of course be collected in any case', but rather to maintain credit for the future. This is quite clearly evident in the practical consequence which a protested bill has in the business world.

relevant paragraph of some law or other. However, the conviction that subject and legal relation have no existence outside the objective norm is just as mistaken theoretically as the conviction that value does not exist and cannot be defined beyond supply and demand, because it is only manifested empirically in price fluctuations.

The presently prevailing mode of legal thought, which gives pride of place to the norm as an authoritatively prescribed rule of conduct, is no less empiricist, and goes hand in hand – as can be observed in economic theories too – with an extreme formalism, devoid of life.

There can be supply and demand for anything you like, including objects which are not products of labour at all. From this it is inferred that value can be determined without reference to the socially necessary labour time required for the production of the object in question. The empirical fact of an individual evaluation is here used as the basis of the formal-logical theory of marginal utility.

In just the same way, the norms decreed by the state can concern the most diverse matters and can exhibit the most varied character. From this it is inferred that the essence of law is exhausted in the norms of conduct, or in the commandment emanating from a higher authority, and that the substance of social relations as such contains no elements especially conducive to the creation of the legal form.

The formal-logical theory of legal positivism is based on the empirical fact that the relations under state protection are the most secure.

To use the terminology of the materialist conception of history, the question for analysis is reduced to the problem of the dialectical relationship between the legal and the political superstructure.

If we acknowledge that the norm is the primary element in every respect, we must first presuppose the existence of a norm-setting authority, of a political organisation in other words, before we start looking for a legal superstructure. In this way, we will inevitably reach the conclusion that the legal superstructure is a consequence of the political superstructure.

However, Marx himself emphasises the fact that the property relation, this most fundamental and lowest layer of the

legal superstructure, stands in such close contact 'with the existing relations of production' that it 'is but a legal expression
for the same thing'.[10] The state, that is, the organisation of
political class dominance, stems from the given relations of
production or property relations. The production relations and
their legal expression form that which Marx, following in
Hegel's footsteps, called civil society.[11] The political superstructure, particularly official statedom, is a secondary, derived
element.

The following quotation illustrates Marx's conception of the
relationship between civil society and the state:

> The egoistic individual in civil society may in his non-
> sensuous imagination and lifeless abstraction inflate himself
> into an atom, i.e., into an unrelated, self-sufficient, wantless,
> absolutely full, blessed being. Unblessed sensuous reality does
> not bother about his imagination, each of his senses compels
> him to believe in the existence of the world and of individuals
> outside him, and even his profane stomach reminds him every
> day that the world outside him is not empty, but is what
> really fills. Every activity and property of his being, every one
> of his vital urges, becomes a need, a necessity, which his self-
> seeking transforms into seeking for other things and human
> beings outside him. But since the need of one individual has no
> self-evident meaning for another egoistic individual capable of
> satisfying that need, and therefore no direct connection with
> its satisfaction, each individual has to create this connection;
> it thus becomes the intermediary between the need of another
> and the objects of this need. Therefore, it is natural necessity,
> the essential human properties however estranged they may
> seem to be, and *interest* that hold the members of civil society
> together; *civil, not political life is their real tie.* It is there-
> fore not the state that holds the atoms of civil society together,
> but the fact that they are atoms only in imagination, in the
> heaven of their fancy, but in reality beings tremendously dif-
> ferent from atoms, in other words, not divine egoists, but
> egoistic human beings. *Only political superstition still imagines*
> *today that civil life must be held together by the state, where-*

---

[10] Marx, 'Preface to the Critique of Political Economy', in Marx and
Engels, *Selected Works*, vol. I, 1969, pp. 503-504. [Ed.]
[11] Ibid., p. 503. [Ed.]

as in reality, on the contrary, the state is held together by civil life.[12]

Marx returns to the same problem in another work, 'Moralising Criticism and Critical Morality', in which he polemicises against a representative of 'true socialism', Karl Heinzen, writing:

> Incidentally, if the bourgeoisie is politically, that is, by its state power, 'maintaining injustice in property relations' (Marx is here quoting Heinzen, E.P.), *it is not creating it.* The 'injustice in property relations' which is determined by the modern division of labour, the modern form of exchange, competition, concentration, etc., *by no means arises from the political rule of the bourgeois class,* but vice versa, the political rule of the bourgeois class arises from these modern relations of production which bourgeois economists proclaim to be necessary and eternal laws.[13]

[12] Marx, 'The Holy Family', in Marx and Engels, *Collected Works,* vol. IV, 1975, pp. 120-121. [Emphases are Pushukanis', which differ considerably from those in the English translation of Marx quoted here. Transl.]

[13] Marx, 'Moralising Criticism and Critical Morality', in Marx and Engels, *Collected Works,* vol. VI, 1976, p. 319. [Emphases are Pashukanis', which differ somewhat from those in the English translation of Marx quoted here. Transl.] It would of course be a gross error to conclude from these statements that there is no place at all for political organisation, and that, in particular, the proletariat has no need to strive to attain state power, this being in no way the most important thing. The syndicalists, who champion *action directe,* make this mistake. The reformists' theory is just as grossly distorted. They have impressed upon their minds the single truth that the political sovereignty of the bourgeoisie originates in the relations of production, and from this they conclude that a violent political revolution led by the proletariat would be impossible and futile. In other words, they transform Marxism into a fatalistic and basically anti-revolutionary doctrine. In reality, of course, the very production relations from which the political sovereignty of the bourgeoisie arises create, in the course of their development, the preconditions for the growth of the political power of the proletariat and, ultimately, for its political victory over the bourgeoisie. It is possible to ignore this dialectic of history by siding consciously or unconsciously, with the bourgeosie against the working class. We shall restrict ourselves here to these brief observations, as our task here is not to refute the mistaken conclusions drawn from the Marxian theory of base and superstructure (all the less so, since this has already been brilliantly dealt with by revolutionary Marxism in the struggle against syndicalism and reformism), but rather to infer from this historical theory certain aspects conducive to the analysis of the structure of law.

According to this, the distance from the production relation to the legal relation is shorter than so-called positive jurisprudence thinks, unable as it is to do without a mediating connecting link, state authority and its norms. The precondition from which economic theory begins is man producing in society. The general theory of law, in so far as it is concerned with fundamental definitions, should start from the same basic prerequisite. Thus the economic relation of exchange must be present for the legal relation of contracts of purchase and sale to arise. Political power can, with the aid of laws, regulate, alter, condition and concretise the form and content of this legal transaction in the most diverse manner. The law can determine in great detail what may be bought and sold, how, under what conditions, and by whom.

From this, dogmatic jurisprudence concludes that all existing aspects of the legal relation, including the subject, are generated by the norm. In reality, the existence of a commodity and money economy is the basic precondition, without which all these concrete norms would have no meaning. Only under this condition does the legal subject have its material base in the person of the subject operating egoistically, whom the law does not create, but finds in existence. Without this base, the corresponding legal relation is *a priori* inconceivable.

The problem becomes clearer still when we consider it at the dynamic and historical level. In this context, we see how the economic relation in its actual workings is the source of the legal relation, which comes into being only at the moment of dispute. It is dispute, conflict of interest, which creates the legal form, the legal superstructure. In the lawsuit, in court proceedings, the economically active subjects first appear in their capacity as parties, that is, as participants in the legal superstructure. Even in its most primitive form, the court is legal superstructure par excellence. The legal differentiates itself from the economic and appears as an autonomous element through legal proceedings. Historically, law begins with dispute, that is, with the lawsuit; only later did it embrace the preceding, purely economic or practical relations, which thus from the very beginning assumed a double-edged economic-legal aspect. Dogmatic jurisprudence forgets this sequence of events, proceeding directly from the end result, from the abstract norms with which the state fills, so to speak, the whole social domain. It acknowledges as legal only those actions which are undertaken within that domain. Corresponding to this over-simplified notion, the fundamental determining factor in the relations of buying-selling, borrowing, lending, and so forth is thought to be, not the actual economic content of these relations, but the imperative addressed to the individual in the name of society. This is the starting point of the practical jurist, which is just as useless for the analysis and explication of the concrete legal system as for the analyis of the legal form in its most general definitions. The state authority introduces clarity and stability into the structure of law, but does not create the premises for it, which are rooted in the material relations of production.

As is well-known, Gumplowicz reaches exactly the opposite conclusion; he proclaims the primacy of the state, that is to say of political power.[14] He turns his attention to the history of Roman law and imagines that he has proved that 'all private law' arose 'as privileges of the ruling class, as public-law preferences', whose purpose was to consolidate power in the hands of the victorious group.

One cannot deny that this theory is persuasive to the extent that it emphasises the element of class struggle and puts an end to idyllic notions as to the origin of private property and state power. Nevertheless, Gumplowicz makes two serious errors. Firstly, he attributes form-creating significance to power in itself, completely overlooking the fact that every social system, even one based on conquest, is conditioned by the state of its social productive forces. Secondly, in speaking of the state, he blurs the distinctions between primitive power relations and 'public authority' in the modern, that is, the bourgeois sense. That is why it follows for him that private law is created by public law. Yet from the fact that the most important institutions of ancient Roman *jus civile* – property, the family, order of inheritance – were created by the ruling class in order to consolidate their power, it is in fact possible to draw a conclusion which is diametrically opposed to Gumplowicz's, namely that 'all state law was once private law'. This would be

14 Ludwik Gumplowicz, *Rechtsstaat und Sozialismus*, Innsbruck, 1881, section 35.

just as true, or rather just as false, since the contradistinction between public and private law corresponds to much more highly developed relations and has little significance when applied to this primitive epoch. If the institutions of *jus civile* do actually represent a mixture of public-law and private-law moments – to use modern terminology – then they also contain to the same extent religious, and, in the wider sense of the word, ritual elements. Consequently, at this stage of development it was not possible to discern the purely legal factor, let alone to express it in a system of general concepts.

The development of law as a system was not predicated on the needs of the prevailing power relations, but on the requirements of trading transactions with peoples who were precisely not yet encompassed within a unified sphere of authority. Moreover, Gumplowicz himself admits this.[15] Trade relations with foreign tribes, with resident aliens (*Perigrini*), plebeians, and in general with any persons not belonging to the public-law association (to use Gumplowicz's terminology), called into being the *jus gentium*, that prototype of the legal superstructure in its purest form. Contrary to the *jus civile* with its cumbersome forms, the *jus gentium* rejects everything which is not connected with the nature and purpose of the economic relation on which it is based. It adapts itself to the nature of this relation, and consequently appears as 'natural' law. It strives to reduce this relation to the fewest possible premises and thus effortlessly becomes a logically harmonious system. Gumplowicz is undoubtedly right to identify specifically juridical logic with the logic of the civil jurist;[16] but he is mistaken in thinking that it was negligence on the part of the state power which enabled the system of private law to develop. His train of thought was roughly this: in view of the fact that private lawsuits did not affect the interests of the state power directly or materially, the latter conceded to the ranks of the jurists complete free-

[15] Ibid., section 36.

[16] This deep inner connection between juridical logic as such and the logic of the civil jurist is indicated also by the historical fact that the general definitions of law developed over a long period as part of the theory of civil law. Only a very superficial consideration of the problem would lead one to believe – as Kavelin does – that this fact can be explained by misconception or misunderstanding. (Cf. Konstantin Dmitrievich Kavelin, *Collected Works, Sobranie sochineniya*, vol. IV, p. 338.)

dom to sharpen their wits in this arena.[17] In the sphere of constitutional law on the other hand, the efforts of the jurists are usually cruelly foiled by reality, for state power does not tolerate any kind of interference in its affairs and does not acknowledge the omnipotence of juridical logic.

It is readily evident that the logic of juridical concepts corresponds to the logic of the social relations of a commodity-producing society. It is precisely in these relations – and not in the permission of authority – that the roots of the system of private law should be sought. Yet the logic of the relations of dominance and subservience can only be partially accommodated within the system of juridical concepts. That is why the juridical conception of the state can never become a theory, but remains always an ideological distortion of the facts.

Thus, wherever we have a primary layer of the legal superstructure, we find that the legal relation is directly generated by the existing social relations of production.

It follows from this that it is unnecessary to start from the concept of the norm as external authoritative command in order to analyse the legal relation in its simplest form. It is sufficient to base the analysis on a legal relation 'whose content is itself determined by the economic relation'[18] and then to investigate the 'statutory' form of this legal relation as a particular case.

The question, raised on the level of actual history, as to whether the norm should be regarded as a prerequisite for the legal relation led us to the problem of the dialectical relationship between the legal and the political superstructure. In the logical and systematic realm, this question is posed as a problem of the relationship between subjective and objective law.

In his textbook on constitutional law, Duguit draws attention to the fact that the same word, 'law', is used to designate things 'which doubtless permeate one another deeply but are, at the same time, very clearly differentiated'.[19] Here he is thinking of law in the objective and the subjective sense.[20] This is indeed one of the most obscure and most disputed points of

17 Gumplowicz, *Rechtsstaat under Sozialismus*, op. cit., section 32.
18 Marx, *Capital*, vol. I, op. cit., p. 178. [Ed.]
19 Leon Duguit, *Etudes de droit public*, 2 vols., Paris, 1901-1903.
20 'Recht': See remarks in the 'Notes to this edition', p. 8. [Ed.]

the general theory of law. We are confronted by a strange dichotomy in the concept whose twin facets, whilst they are to be found on different planes, undoubtedly complement one another. Law is simultaneously the form of external authoritarian regulation and the form of subjective private autonomy. In the one case, the fundamental, substantive characteristic is that of unconditional obligation, of absolute external coercion, while, in the other, it is the characteristic of freedom, guaranteed and recognised within certain limits. Law appears sometimes as a principle of social organisation, and at other times as a means of enabling individuals to 'define themselves within society'. On the one hand, law merges completely with the external authority, while on the other, it is just as completely opposed to every external authority which does not acknowledge it. Law as a synonym for official statedom, and law as the watchword of revolutionary struggle: this is the field of endless controversies and of the most unimaginable confusion.

Acknowledging the profound contradiction implied here has led to many attempts to eliminate somehow this unwelcome cleavage of concepts, to sacrifice one of these two 'meanings' to the other. Thus the same Duguit, who, in his textbook, labels the expressions 'subjective law and objective law' as 'apposite, clear and exact', uses all his ingenuity in another work to prove that subjective law is simply based on a misconception, *'a metaphysical conception which is untenable in an age of realism and positivism such as ours'.*[21]

The opposite tendency, represented in Germany by Bierling and in Russia by the psychologists headed by Petrazhitsky, tends on the contrary to portray objective law as an 'emotional projection' devoid of any real significance, as a phantasm, as the externalisation of inner psychological processes, and so on.[22]

[21] Duguit, *Les transformations du droit public*, Paris 1913; translated by F. and H. Laski, as *Law in the Modern State*, London: G. Allen & Unwin, 1921; New York: Howard Fertig, 1970. [Transl.]

[22] Cf. in Bierling: 'Corresponding with a general inclination of the human mind, we think of law as first and foremost something objective, something which actually exists, something standing above the professional jurist. Then too, this undoubtedly has its practical merit. Only one must not forget meanwhile that this 'objective law' always remains – even when it has assumed an external form peculiar to it in written law – merely a form of *our conception* of law. Law itself actually exists, like every other product of mental activity, only in the mind, especially in

We shall bypass the psychological school and related tendencies for the moment and concern ourselves with the notion that law should be grasped exclusively as an objective norm.

If one starts from this conception one has, on the one hand, the authoritative, prescriptive *Ought* as a norm and on the other, the subjective obligation created in conformity with this regulation.

The dichotomy appears to have been completely eradicated; however this suppression is only apparent, for no sooner does one come to apply this formula than attempts are renewed to reintroduce, by subterfuge, all the nuances which are indispensable to the concept of 'subjective law'. We are confronted once again by the same two facets, only with the difference that one of the two, namely subjective law, is by all manner of subterfuge portrayed as a kind of shadow; for there is no possible combination of commands and obligations which would give us subjective law in the autonomous and perfectly real significance through which it is personified by any one property-owner in bourgeois society. Indeed, taking property as one's example makes this perfectly clear. For if the attempt to reduce property *law* to *prohibitions* against third parties is no more than a logical trick, a garbled construct turned inside out, then the portrayal of bourgeois property right as social obligation is hypocrisy to boot.[23]

the minds of the professional jurists'. (Ernst Rudolf Bierling, *Juristische Prinzipienlehre*, 5 vols., Freiburg i.B. and Leipzig, 1894, vol. I, p. 145.)
[23] In his commentary on the Civil Code of the RSFSR, Goichbarg emphasises that progressive bourgeois jurists are already ceasing to consider private property as an arbitrary subjective right, seeing it instead as an asset placed at the disposal of the individual [and associated with positive obligations to the whole – in 3rd Russian ed., but omitted in German ed.—Ed.] Goichbarg refers specifically to Duguit, who maintains that the owner of capital should be legally protected only because, and to the extent that, by correctly investing his capital, he fulfils a socially useful function.

Such observations, typical of bourgeois jurists, are the early warning signals of the downfall of the capitalist epoch. Yet the bourgeoisie concedes such observations on the social functions of property only because they do not seriously commit it to anything. For the real opposite of property is not property *understood* as a social function, but socialist planned economy, that is the abolition of property. The significance of private property, its subjectivism, does not lie in the fact that each person 'eats his *own* bread', not, that is, in the act of individual consumption, even if this be productive consumption. Rather, it lies in

Any property-owner, as well as everyone around him, understands very well that the *right to which he is entitled* as a property-owner has only so much in common with obligation: that it is its polar opposite. Subjective law is the primary law, for it is based, after all, on material interest, which exists independently of the external, or conscious, regulation of social life.

The subject as representative and addressee of every possible claim, the succession of subjects linked together by claims on each other, is the fundamental legal fabric which corresponds to the economic fabric, that is, to the production relations of a society based on division of labour and exchange.

A social organisation with the means of coercion at its disposal is the concrete totality which we must arrive at after first comprehending the legal relation in its purest and simplest form. Consequently, when analysing the legal form, obligation, stemming from an imperative or a command, seems to be a factor which concretises and complicates things. Legal obligation in its most abstract, simple, form should be seen as the

---

circulation, in the acts of appropriation and alienation, in the exchange of commodities, in which the social economic purpose is simply a blind outcome of private purposes and private autonomous decisions.

Duguit's declaration that the property-owner should only be protected when he fulfils his social obligation has no meaning in this general form. It is hypocrisy in the bourgeois state, and obscures the facts in the proletarian state. For if the proletarian state were able to refer every property-owner directly to his social function, it would achieve this by stripping him of the right to dispose over his property. However, if it is *economically* not in a position to do this, it will be compelled to protect private interests as such, merely imposing certain quantitative limits on it. It would be an illusion to maintain that anyone within the borders of the Soviet Union who has accumulated a certain quantity of ten-rouble notes is protected by our laws and courts for the sole reason that this person has found, or will find, a socially necessary use for the amassed notes. Further, Goichbarg appears to have completely overlooked property in capital in its most abstract, money, form, and makes his observations as if capital only existed in the concrete form of productive capital. The anti-social aspects of private property can only be eliminated *de facto*, that is by the development of socialist planned economy at the expense of the market economy. But there is absolutely no formula, be it even drawn from the writings of the most progressive Western European jurists imaginable, which can transform the legal transactions arising out of our Civil Code into socially useful transactions, and can transform every property-owner into a person performing social functions. Such an abolition of private economy and private law on paper can only serve to obscure the perspective of its real abolition.

reflection and correlate of the subjective legal claim. In analysing the legal relation, it is abundantly clear that obligation cannot exhaust the logical content of the legal form. Indeed, it is not even an autonomous element of it. Obligation always figures as the reflection and correlate of right. One party's debt is something owed and guaranteed to the other party. Right, from the creditor's point of view, is obligation for the debtor. The category of law only becomes logically complete where it embraces the representative and bearer of right, whose rights are nothing but the obligations of others towards him. This dualistic nature of law is particularly emphasised by Petrazhitsky, who gives it a fairly flimsy basis in the form of the psychological theory he invents *ad hoc*. However, it should be noted that this dialectical relationship of right and obligation has been very precisely formulated by other jurists, free of any psychologism.[24]

Thus the legal relation not only shows us law in its actual workings, but also reveals the characteristic traits of law as a logical category. In contrast to this, the norm as such, as a prescriptive *Ought*, is as much an element of ethics, aesthetics, and technology as it is of law.

Contrary to Alexeyev's opinion, the difference between technology and law in no way resides in the fact that technology has a purpose external to its subject-matter whilst in the legal system every purpose must be posited as an end in itself.[25] In what follows, it will be argued that the only 'end in itself' for the legal system is commodity circulation. Let us look at the technique of a teacher or a surgeon: for the teacher the child's psyche is his subject, while for the surgeon it is the organism of the person being operated on. But no-one, presumably, would dispute the fact that the subject is also the purpose in these cases.

The legal system differs from every other form of social system precisely in that it deals with private, isolated subjects. The legal norm acquires its *differentia specifica*, marking it out from the general mass of ethical, aesthetic, utilitarian, and other

[24] Cf. for example Adolf Merkel, *Juristische Enzyklopädie*, Berlin and Leipzig, 1884, section 146; and Nikolay Mikhaylovich Korkunov, *Encylopaedia of Law (Entsiklopediya prava)*.
[25] I. Alexeyev, *Introduction to the Study of Law (Vvedenie v izuchenie prava)*, Moscow, 1918, p. 114.

such regulations, precisely because it presupposes a person endowed with rights on the basis of which he actively makes claims.[26]

The attempt to make the idea of external regulation the fundamental logical element in law leads to law being equated with a social order established in an authoritarian manner. This current in juridical thought accurately reflects the spirit of the age in which the Manchester school and free competition were superseded by the monopolies of large-scale capital and by imperialist policies.

Finance capital sets much greater store on a strong arm and discipline than on the 'eternal and inalienable rights of the individual citizen'. The capitalist proprietor, transformed into the recipient of dividends and profits on speculations, cannot but regard the 'sacred right of property' with a certain cynicism. One need only refer to Ihering's diverting lamentations about the 'mire of speculation and fraudulent stockjobbery', beneath which the 'healthy feeling for law' has been submerged.[27]

It is not difficult to establish that the idea of unconditional subjection to an external norm-setting authority has nothing whatever to do with the legal form. One need only look at examples of structures which take such a conception to the limit, and are therefore particularly clear. Take the military unit set out in formation, where many people are subordinate in their movements to a common order in which the only active and autonomous principle is the will of the commander. Or take another example: the Jesuit order, in which all brothers of the order carry out the will of the superior blindly and without protest. One has merely to immerse oneself in these examples to reach the conclusion that the more consistently the principle of authoritarian regulation is applied, excluding all reference to separate autonomous wills, the less ground there remains for applying the category of law. This is particularly noticeable in the realm of so-called public law. It is here that legal theory encounters the most serious difficulties. Generally speaking, the very phenomenon that Marx characterised as the

---

[26] 'Law is not given to the person in need of it for nothing' (M. A. Muromtsev, *The Formation of Law, Obrazovanie prava*, 1885, p. 33).
[27] Rudolf von Ihering, *Der Kampf um's Recht* (2nd ed., Regensburg, 1872), Vienna, 1900.

separation of the political state from civil society is reflected in the general theory of law as two independent problems, each of which has a particular place in the system and is resolved independently of the other. The first problem is purely abstract: it is the cleavage of the basic concept into the two facets which we have described above. Subjective law is the characteristic trait of the egoistic person as 'a member of civil society, an individual withdrawn into himself, into the confines of his private interests and private caprice, and separated from the community'.[28] Objective law is the expression of the bourgeois state as a whole, which 'feels itself to be a political state and asserts its universality only in opposition to these elements of its being'.[29]

The problem of subjective and objective law is the general philosophical formulation of the problem of the individual as member of civil society and as citizen of a state. This problem comes up yet again (this time in more concrete form) as the problem of public and private law. The problem is here reduced to delineating various spheres of law which actually exist, to distributing under various headings institutions which emerged historically. Dogmatic jurisprudence, with its formal-logical method, can obviously solve neither the first nor the second problem, nor can it elucidate the connection between the two.

The split into public and private law presents specific difficulties for this reason alone: that only in the abstract can one draw a line between the egoistic interest of man as a member of civil society and the abstract universal interest of the political whole. In reality, these elements are interdependent, so that it is impossible to indicate the particular legal institutions which embody this much-trumpeted private interest entirely and in pure form.

Another difficulty is that even if the jurist manages to draw an empirical line between the institutions of public and private law more or less successfully, he comes up against the identical problem again within the limits of each of these two domains. This problem, which had to all appearances been solved, appears this time in a somewhat different, abstract,

[28] Marx, 'On the Jewish Question', in Marx and Engels, *Collected Works*, vol. III, 1975, p. 164. [Ed.]
[29] Ibid, p. 153. [Ed.]

formulation as the contradiction between subjective and objective law. Subjective public rights – just the same old private rights again (and thus private interests too), resurrected and merely somewhat metamorphosed – intrude into a sphere in which the impersonal common interest, reflected in the norms of objective law, should hold sway. Yet while civil law, which is concerned with the fundamental, primary level of law, makes use of the concept of subjective rights with complete assurance, application of this concept in public-law theory creates misunderstandings and contradictions at every step. For this reason, the system of civil law is distinguished by its simplicity, clarity and perfection, while theories of constitutional law teem with far-fetched constructs which are so one-sided as to become grotesque. The form of law with its aspect of subjective right (*Berechtigung*)[30] is born in a society of isolated bearers of private egoistic interests. If all economic life is to be built on the principle of agreement between autonomous wills, every social function, in reflecting this, assumes a legal character. It is in the nature of political organisation that it does not promote the full development of, nor does it have such paramount significance for, private interests as does the economic system of bourgeois society. Hence subjective public rights also appear as something ephemeral, lacking genuine roots, something eternally dubious. Yet at the same time the state is not a legal superstructure, but can only be *conceived of* as such.[31]

The theory of law cannot equate the rights of the legislature, those of the executive, and so on with, for example, the right of the creditor to restitution of the sum borrowed from him. This would imply substituting isolated private interest for the dominance of the universal, impersonal interest of state assumed by bourgeois ideology. But at the same time, every jurist is aware of the fact that he cannot endow these rights with any other content in principle without the legal form slipping from his grasp altogether. Constitutional law is only able to

[30] For Kelsen's explanation of the term *Berechtigung* the reader is referred to note 22 of the introduction by Pashukanis (p. 58). [Ed.]
[31] 'For juridical knowledge it is exclusively a matter of answering the question: how should one conceive of the state in legal terms?' (Georg Jellinek, *System der subjektiven öffentlichen Rechte* (Freiburg i.B., 1892), Tübingen, 1905, p. 13).

exist as a reflection of the private-law form in the sphere of political organisation, otherwise it ceases to be law entirely. Every attempt to present a social function as it is, simply as a social function, and a norm simply as an organisational regulation, would mean the death of the legal form. The real prerequisite for such an abolition of the legal form and of legal ideology is, however, a society in which the contradiction between individual and social interests has been broken down.

But the very thing which characterises bourgeois society is that universal interests are disengaged from, and set in opposition to, private interests. In this antithesis, they themselves involuntarily assume the form of private interests, that is, legal form. As could be expected, the juridical factors in state organisation are primarily those which can be adapted to the framework of conflicting isolated private interests.[32]

Goichbarg even disputes the necessity for separating out the concepts of public and private law:

[32] Cf. for example Kotlyarevsky's observations on the right to vote: 'In the constitutional state, the elector fulfils a specific function imposed on him by the state order laid down in the constitution. But it is impossible simply to attribute this function to the elector while ignoring the right it confers on him'. For our part, we would like to add that it is equally impossible simply to transform bourgeois property into a social function. Kotlyarevsky is quite right to give further emphasis to the fact that, if one follows Labande in denying the element of the elector's subjective right, 'the representatives' eligibility for election loses all legal significance, becoming purely a question of technical expediency'. Here too we find, yet again, the same contradiction between technical expediency, based on unity of purpose, and legal organisation, based on the differentiation and opposition of private interests. Finally, representative constitutional government is ultimately stamped as legal with the introduction of the judicial, or judicial-administrative protection of the elector's rights. The judicial process and the conflict of parties figures here too as an important element in the legal superstructure. (Cf. S. A. Kotlyarevsky, *Power and Law, Vlast' i pravo*, Moscow, 1915, p. 25.)

Constitutional law first becomes the subject of juridical attention as constitutional law with the emergence of mutually warring forces, like king and parliament, upper and lower house, the executive and the people's elected representatives. The same is true of administrative law. Its legal content consists solely in safeguarding the rights of the population on the one hand, and the representatives of the bureaucratic hierarchy on the other. Over and above this, administrative law, or police law, as it used to be called, displays a motley mixture of technical regulations, political prescriptions, and so on.

The jurists have never been successful in dividing law into public and private law . . . , and this division is only recognised nowadays by the most reactionary jurists, ours included.[33]

Goichbarg further substantiates this notion that it is unnecessary to divide law up into public and private law with several observations. First, that the Manchester-school principle of non-intervention in economic affairs by the state is outdated in the twentieth century. Second, that unlimited individual choice in the economic sphere is detrimental to the interests of the whole. Third, that even in countries where there has not been a proletarian revolution, there are numerous structures which combine the realms of public and private law. Lastly, that in our country, where economic activity is, in the main, concentrated in the hands of state agencies, there is no longer any point in differentiating the concept of private law from other concepts.

It would seem that this line of argument rests on a whole series of misconceptions. The choice of one or other direction in practical politics does not decide anything with reference to the theoretical difference between certain concepts. We can be sure, for example, that the establishment of economic relations based on market transactions has many detrimental consequences. Nevertheless, it does not follow from this that the distinction between the concepts 'use value' and 'exchange value' is theoretically untenable. Neither would there be any point in reiterating the assertion that the spheres of public and private law merge, if one were unable to distinguish between these two concepts. How could entities which have no separate existence merge? Goichbarg's objections are based on the notion that the abstractions mentioned, 'public law' and 'private law', are not the product of historical development, but were simply dreamt up by the jurists. Yet it is precisely this contradiction which is the most typical attribute of the legal form as such. The division of law into public and private law characterises this form from the historical as well as from the logical aspect. Positing this contradiction as simply non-existent in no way makes one superior to the 'reactionary' practical jurists; scholastic definitions with which they operate.

[33] A. Goichbarg, *Economic Law* (*Khozyaystvennoe pravo*), p. 5.

on the contrary, it forces one to employ the same formal, scholastic definitions with which they operate.

Thus the very concept of 'public law' can only be developed through its workings, in which it is continually repulsed by private law; so much so that it attempts to define itself as the antithesis of private law, to which it returns, however, as to its centre of gravity.

The attempt to take the opposite course, to locate the fundamental definitions of private law – which are none other than the definitions of law itself – by starting out from the concept of the norm, can produce no result other than lifeless formal constructs with ineradicable internal contradictions. Law as a function ceases to be law, while legal power (Berechtigung) without the supporting private interest becomes something intangible and abstract which can easily turn into its own opposite, that is, into obligation. (Every public right is of course simultaneously obligation.) Parliament's 'right' to approve the budget is as unstable, problematic, and in need of interpretation as the 'creditor's right' to recover the sum lent by him is easily understood and 'natural'. Whereas in civil law disputes are dealt with on the level Ihering called symptomatic of law, here the basis of jurisprudence itself is in question. Herein lies the source of the methodological wavering and uncertainty which threaten to transform jurisprudence sometimes into sociology, sometimes into psychology.

On the basis of the above exposition, several of my critics, among them Razumovsky and Il'insky, apparently thought that I had set myself the task of 'constructing a theory of pure jurisprudence'. Il'insky concluded that this task had remained unfulfilled:

> The author has produced a theory of law which is basically sociological, although he intended to construe it as pure jurisprudence.[34]

In Razumovsky's case, he expresses no definite opinion about my results, but neither does he question the existence of the above-mentioned intention, which he censures most severely:

> His (that is mine, E.P.) concern that methodological analyses could transform jurisprudence into sociology or psychology

[34] Cf. T. Il'insky, in *Molodaya Gvardiya*, no. 6.

merely illustrates his inadequate conception of the nature of Marxist analysis . . . This is all the more mysterious – (thus my critic expresses his astonishment, E.P.) – since Pashukanis himself sees a certain discrepancy between sociological and legal truth and since he knows that the legal view is one-sided.[35]

Yes, it is indeed strange. On the one hand I am presumed to be afraid that jurisprudence will be transformed into sociology, on the other I admit that the legal view is one-sided. On the one hand I try to present a theory of pure jurisprudence, while on the other it turns out that I have produced a sociological theory of law. How can these contradictions be resolved? The solution is very simple. As a Marxist, I did not set myself the task of constructing a theory of pure jurisprudence, nor could I set myself such a task. From the very first I had a clear idea of the goal which Il'insky says I have reached unconsciously. My aim was this: to present a sociological interpretation of the legal form and of the specific categories which express it. This is exactly why I gave my book the explanatory subtitle: *Contribution to a Critique of the Fundamental Juridical Concepts*. Indeed my task would have been quite pointless, were I to deny the existence of the legal form itself and to discard the categories which express this form as idle fancies.

When I describe the unreliability and inadequacy of juridical constructs in the field of public law, and, in the process, speak of the methodological vacillations and uncertainty which threaten to transform jurisprudence sometimes into sociology, at others into psychology, it is a strange idea to imagine that I mean thereby to warn against any attempt to carry out a sociological critique of jurisprudence from a Marxist standpoint. After all, such a warning would be directed primarily at myself. The propositions which elicited Razumovsky's amazement, and which he explains by my inadequate understanding of the nature of Marxist analysis, refer to the conclusions of bourgeois jurisprudence, whose conceptual structure loses all credibility as soon as it loses sight of the exchange relation (in the widest sense of the word). Perhaps I should have made it obvious by

[35] Cf. Isaak Petrovich Razumovsky, in *Vestnik Kommunisticheskoy Akademii*, no. 8.

an explicit reference that that phrase about the 'danger threatening jurisprudence' is an allusion to the lament of a bourgeois legal philosopher. This lament relates, not to the Marxist critique (which had not yet ruffled the minds of the 'pure jurists' in those days), but to the attempts of bourgeois jurisprudence itself to mask the limitations of its own method by borrowing from sociology and psychology. So far was it from my mind to imagine that I could be seen as a 'pure jurist', whose heart grieves for jurisprudence threatened by the Marxist critique, that I did not take such precautionary measures.

4. Commodity and Subject

Every legal relation is a relation between subjects. The subject is the atom of legal theory, its simplest, irreducible element. Consequently we begin our analysis with the subject.

Razumovsky does not agree with me that the analysis of the concept 'subject' is basic to the investigation of the legal form. This category of developed bourgeois society seems to him firstly, too complicated, and secondly, not characteristic of preceding historical epochs. In his opinion, 'the development of the relation which is fundamental to all class society' should be the point of departure',[1] and this, as Marx says in his *Introduction*, is 'property, which develops from actual appropriation into legal property'.[2] In laying bare the course of this development, Razumovsky too concludes that it is only in the process of development that private property in the modern sense takes shape as such, and then only to the extent that it goes hand in hand, not only with the 'possibility of unimpeded possession thereof', but also with the 'possibility of its alienation'.[3] In effect, this also means that the legal form in its developed state corresponds to bourgeois-capitalist social relations. It is clear that particular forms of social relations do not invalidate either these relations in themselves, or the laws upon which they are founded. Thus the appropriation of a product within a given social formation and thanks to the forces at work within it is a basic fact or, if you like, a basic law. But this relation of private property assumes legal form only at a certain stage of develop-

[1] Isaak Petrovich Razumovsky, *Problems of the Marxist Theory of Law* (*Problemy marksistskoy teorii prava*), Moscow, 1925, p. 18.
[2] Cf. Marx, *Grundrisse: Introduction to the Critique of Political Economy*, 1973 ed., p. 102. [Ed.]
[3] Razumovsky, *Problems . . .*, op. cit., p. 114.

ment of the productive forces and of the corresponding division of labour. Razumovsky thinks that by basing my analysis on the concept of the subject, I am eliminating the relationship of dominance and subservience from my investigation, whereas, of course, possession and property are inextricably bound up with this relationship. It would not occur to me to dispute this link. I merely maintain that property becomes the basis of the legal form only when it becomes something which can be freely disposed of in the market. The category of the subject serves precisely as the most general expression of this freedom. What, for example, is the significance of legal ownership of land and soil? 'Simply', says Marx, 'that the landowner can do with his land what every owner of commodities can do with his commodities'.[4] Yet it is precisely by completely liberating land-ownership from the relation of dominance and subservience that capitalism transforms feudal land-ownership into modern landed property. The slave is totally subservient to his master. This is exactly why this exploitative relationship requires no specifically legal formulation. The wage worker, on the contrary, enters the market as a free vendor of his labour power, which is why the relation of capitalist exploitation is mediated through the form of the contract. I think these examples suffice to demonstrate the decisive importance of the subject for any analysis of the legal form.

Idealist theories of law develop the concept of the subject from this or that general idea, that is, by purely speculative means.

> The fundamental concept of law is freedom . . . The abstract concept of freedom is the possibility of self-determination . . . Man is the subject of rights for the reason that he has this possibility, and that he has free-will.[5]

Compare Hegel:

> Personality essentially involves the capacity for rights and constitutes the concept and the basis (itself abstract) of the

---

[4] Marx, *Capital*, vol. III, 1962 ed., p. 602.
[5] Georg Friedrich Puchta, *Kursus der Institutionen*, 3 vols., Leipzig, vol. I, 1850, pp. 4-9. [Vol. I has a second title page reading: *Einleitung in die Rechtswissenschaft und Geschichte des Rechts bey dem römischen Volk*. Transl.]

system of abstract and therefore formal right. Hence the imperative of right is: 'Be a person and respect others as persons'.[6]

And further:

What is immediately different from free mind is that which, both for mind and in itself, is the external pure and simple, a thing, something not free, not personal, without rights.[7]

In what follows, we shall see in what sense this contradistinction between thing and subject provides the key to an understanding of the form of law. In contrast, dogmatic jurisprudence employs this concept from the formal aspect. In its eyes, the subject is nothing more than a 'means of juridical qualification of a phenomenon from the point of view of its fitness or unfitness for participation in legal intercourse'.[8] Hence dogmatic jurisprudence avoids altogther putting the question of how man was transformed from a zoological individual into a legal subject, since it proceeds from the legal process as from a finished form, given a priori.

In contrast to this, Marxist theory considers every social form historically. Consequently, it sets itself the task of elucidating those historically given material conditions which brought this or that category into being. The material preconditions for the community of law or for transactions between legal subjects are specified by Marx himself in volume one of Capital, albeit only in passing, in the form of fairly general allusions. Nonetheless, these hints contribute far more to the understanding of the juridical element in human relations than any of those bulky treatises on the general theory of law. In Marx, the analysis of the form of the subject follows directly from the analysis of the commodity form. Capitalism is a society of commodity-owners first and foremost. This means that social relations in the production process assume a reified form in that the products of labour are related to each other as values. The commodity is a thing in which the concrete multi-

---

[6] Hegel, *Philosophy of Right*, translated by T. M. Knox, Oxford: Oxford University Press, 1957, section 36.
[7] Ibid., section 42.
[8] Cf. Rozhdestvensky, *Theory of Subjective Public Rights* (*Teoriya sub'yektivnykh publichnykh prav*), p. 6.

plicity of use-values becomes simply the material shell of the abstract property of value, which manifests itself as the capacity to be exchanged with other commodities in a specific relation.

This property appears as an intrinsic natural property of objects themselves, according to a sort of natural law which operates behind people's back, quite independently of their will.

Whereas the commodity acquires its value independently of the will of the producing subject, the realisation of its value in the process of exchange presupposes a conscious act of will on the part of the owner of the commodity, or as Marx says:

> Commodities cannot themselves go to market and perform exchange in their own right. We must, therefore, have recourse to their guardians, who are the possessors of commodities. Commodities are things, and therefore lack the power to resist man. If they are unwilling, he can use force; in other words, he can take possession of them.[9]

It follows that the necessary condition for the realisation of the social link between people in the production process – reified in the products of labour and disguised as an elementary category (*Gesetzmässigkeit*) – is a particular relationship between people with products at their disposal, or subjects whose 'will resides in those objects'.[10]

> ... That goods contain labour is one of their intrinsic qualities; that they are exchangeable is a distinct quality, one solely dependent on the will of the possessor, and one which presupposes that they are owned and alienable.[11]

At the same time, therefore, that the product of labour becomes a commodity and a bearer of value, man acquires the capacity to be a legal subject and a bearer of rights.[12]

9 Marx, *Capital*, vol. I, 1976 ed., p. 178.
10 Ibid.
11 Rudolf Hilferding, *Böhm-Bawerk's Criticism of Marx*, edited with an introduction by P. M. Sweezy, London: Merlin Press, 1975, pp. 187-188.
12 Man as a commodity, in other words the slave, becomes a reflected subject as soon as he acts as someone disposing over commodities (objects) and participates in circulation. (On the rights of slaves to con-

The person whose will is declared as decisive is the legal subject.[13]

Simultaneously with this, social life disintegrates, on the one hand into a totality of spontaneously arising reified relations (including all economic relations: price level, rate of surplus value, profit rates and so forth) – in other words, the kind of relations in which people have no greater significance than objects – and, on the other hand, into relations of a kind where man is defined only by contrast with an object, that is, as a subject. The latter exactly describes the legal relation. These are the two basic forms, which differ from one another in principle, but are at the same time interdependent and extremely closely linked. The social relation which is rooted in production presents itself simultaneously in two absurd forms: as the value of commodities, and as man's capacity to be the subject of rights.

Just as in the commodity, the multiplicity of use-values natural to a product appears simply as the shell of value, and the concrete types of human labour are dissolved into abstract human labour as the creator of value, so also the concrete multiplicity of the relations between man and objects manifests itself as the abstract will of the owner. All concrete peculiarities which distinguish one representative of the *genus homo sapiens* from another dissolve into the abstraction of man in general, man as a legal subject.

If objects dominate man economically because, as commodities, they embody a social relation which is not subordinate to man, then man rules over things legally, because, in his capacity as possessor and proprietor, he is simply the personification of the abstract, impersonal, legal subject, the pure product of social relations; in Marx's words:

---

clude legal transactions under Roman law, see Yosif Alekseyevich Pokrovsky, *History of Roman Law, Istoriya rimskogo prava*, 2nd ed., Petrograd, 1915, vol. II, p. 294.) In contrast to this, in modern society, the free man, that is to say the proletarian, when seeking in this role a market for the sale of his labour power, is treated as an object and is subject to the same prohibitions and quota allocations under the immigration laws as are other commodities imported across national boundaries.

[13] Bernhard Windscheid, *Lehrbuch des Pandektenrechts*, 9th ed., 3 vols., Frankfurt a.M., 1906, vol. I, section 49.

In order that these objects may enter into relations with each other as commodities, their guardians must place themselves in relation to one another as persons whose will resides in those objects, and must behave in such a way that each does not appropriate the commodity of the other, and alienate his own, except through an act to which both parties consent. The guardians must therefore recognise each other as owners of private property.[14]

Obviously the historical evolution of property as an institution of law, with the many and varied methods of appropriating and protecting it, with all its modifications in relation to different objects and so on, took place in a far less well-ordered and consistent manner than the logical deduction set out above might suggest. Yet this deduction alone reveals the universal significance of the historical process.

After he has become slavishly dependent on economic relations, which arise behind his back in the shape of the law of value, the economically active subject – now as a legal subject – acquires, in compensation as it were, a rare gift: a will, juridically constituted, which makes him absolutely free and equal to other owners of commodities like himself. 'Everyone shall be free, and shall respect the freedom of others . . . Everyone possesses *his own* body as the free tool of his will'.[15]

This is the premise from which the natural-law theoreticians start. The idea of the isolated and self-contained nature of the human personality, this 'natural state' from which the 'conflict of freedom to infinity' follows, corresponds exactly to commodity production, where the producers are formally autonomous, linked only by the artificially created legal system. This legal condition itself or, to quote Fichte again, 'the co-existence of many free beings, who shall all be free, the freedom of one not interfering with the freedom of others', is nothing but the idealised market, transported to the nebulous heights of philosophical abstraction, liberated from crude empiricism. This idealised market is where the autonomous producers come together; for, as another philosopher tells us:

---

14 Marx, *Capital*, vol. I, ed. cit., p. 178.
15 Johann Gottfried Fichte, *Rechtslehre*, Leipzig, 1912, p. 10; [cf. English translation, *The Science of Rights*, translated by A. E. Kroeger, Philadelphia: J. B. Lippincott & Co., 1869, pp. 19, 94 and 178.—Ed.]

in commercial transactions, both parties do as they wish, taking no greater liberty than they themselves grant others.[16]

The increasing division of labour, improvements in communications, and the resulting development of exchange, made value an economic category, that is to say, the embodiment of supra-individual social relations of production. For this to occur, separate casual acts of exchange must be transformed into expanded, systematic commodity circulation. At this stage of development, value ceases to be casual appraisal, loses the quality of a phenomenon of the individual psyche, and acquires objective economic significance. In the same way, there are real conditions necessary for man to be transformed from a zoological individual into an abstract, impersonal legal subject, into the legal person. These real conditions are the consolidation of social ties and the growing force of social organisation, that is, of organisation into classes, which culminates in the 'well-ordered' bourgeois state. At this point the capacity to be a legal subject is definitively separated from the living concrete personality, ceasing to be a function of its effective conscious will and becoming a purely social function. The capacity to act is itself abstracted from the capacity to possess rights. The legal subject acquires a double in the shape of a representative, and himself attains the significance of a mathematical point, a centre in which a certain number of rights is concentrated.

As a result, bourgeois-capitalist property ceases to be unstable, precarious, purely factual property which may at any moment be contested and have to be defended, weapon in hand. It is transformed into an absolute, fixed right which follows the object wherever chance may take it, and which, ever since bourgeois civilisation extended its rule to encompass the whole globe, has been protected the world over by laws, police and lawcourts.[17]

[16] Herbert Spencer, *Social Statics*, 1850, Chapter 13.
[17] The development of so-called rules of warfare is nothing but a gradual consolidation of the inviolability of bourgeois property. Until the French Revolution, populations were plundered by their own as well as by enemy soldiers, without hindrance or prohibition. Benjamin Franklin was the first to proclaim as a political principle, in 1785, that in future wars: 'peasants, labourers and merchants should be able to follow their occupations peacefully, under the protection of both warring parties'. In his *Contrat Social*, Rousseau posits the rule that war would

At this stage of development, the so-called 'will-theory' of subjective rights begins to seem out of touch with reality.[18] People prefer to define law in the subjective sense, as the 'share of earthly possessions which the general will guarantees one person as his acknowledged due'. This entirely precludes the necessity for the person to exercise his will, or to have the capacity to act. Dernburg's definition is indeed better suited to the mental framework of the modern jurist who deals with the legal capacity of idiots, babes in arms, legal persons, and so forth. In contrast to this, the ultimate consequences of the will-theory are synonymous with the exclusion of the categories cited from the ranks of legal subjects.[19] There is no doubt that Dernburg comes closer to the truth when he conceives of the legal subject as an exclusively social phenomenon. Yet it is perfectly clear to us why the element of will plays such a decisive role in constructing the concept of the legal subject. Dernburg does concede this in part, when he asserts:

> Rights in the subjective sense existed historically long before the formation of a political system conscious of itself. They were based in the personality of the individual and the respect for his person and property which he was able to exact and enforce. Only by abstraction from the contemplation of existing subjective rights did the concept of the legal system gradually take shape. To hold that rights in the subjective

be waged between states, but not between the citizens of those states. The legislation of the [Geneva] Convention punished soldiers very harshly for plundering, whether in their own or in a foreign country. It was not until 1899, in the Hague, that the principles of the French Revolution were given the status of statutes of international law. (Moreover, justice demands reference here to the fact that, wheeras Napoleon felt some embarrassment when imposing the Continental blockade, and saw fit to justify this measure in his message to the Senate as a measure 'which affects the interests of private persons as a result of the conflicts of rulers' and 'is reminiscent of the barbarity of long-gone centuries'; yet, in the last World War, the bourgeois governments injured the property rights of the citizens of both warring parties openly and without a trace of embarrassment).

[18] Cf. Heinrich Dernburg, *Pandekten*, 7th ed., 3 vols., Berlin, 1902, vol. I, section 39.

[19] In relation to legal persons, see Alois von Brinz, *Lehrbuch der Pandekten* (2 parts, Erlangen, 1857-71; 2nd ed., 4 vols., Erlangen, 1873-95), vol. II, p. 984.

sense merely emanate from law in the objective sense is therefore ahistorical and incorrect.[20]

Obviously only someone who had not merely a will at his disposal, but also wielded a considerable degree of power, was able to 'exact and enforce'. Like the majority of jurists, Dernburg tends to treat the legal subject as 'personality in general', that is to say, as an eternal category beyond particular historical conditions. From this point of view, being a legal subject is a quality inherent in man as an animate being graced with a rational will. In reality, of course, the category of the legal subject is abstracted from the act of exchange taking place in the market. It is precisely in the act of exchange that man puts into practice the formal freedom of self-determination. The market relation provides a specifically legal illustration of the contradiction between subject and object. The object is the commodity, the subject is the owner of the commodity, who disposes of it in the acts of appropriation and alienation. It is in the exchange transaction in particular that the subject figures for the first time in all the fullness of its definitions. The formally more perfected concept of the subject, which retains only legal capacity, tends rather to distance us from the true historical significance of this juridical category. That is why the jurists find it so difficult to do without the element of active will in the concepts of 'subject' and 'subjective law'.

The sphere of dominance which has taken on the form of subjective law is a social phenomenon attributed to the individual in the same way that value – likewise a social phenomenon – is attributed to the object as a product of labour. Legal fetishism complements commodity fetishism.

Hence, at a particular stage of development, the social relations of production assume a doubly mysterious form. On the one hand they appear as relations between things (commodities), and on the other, as relations between the wills of autonomous entities equal to each other – of legal subjects. In addition to the mystical quality of value, there appears a no less enigmatic phenomenon: law. A homogeneously integrated relation assumes two fundamental abstract aspects at the same time: an economic and a legal aspect.

[20] Dernburg, *Pandekten*, op. cit., vol. I, section 39.

In the development of juridical categories, the ability to perform exchange transactions is only one of various concrete manifestations of the general capacity to act and of legal capacity. Historically, however, it was precisely the exchange transaction which generated the idea of the subject as the bearer of every imaginable legal claim. Only in commodity production does the abstract legal form see the light; in other words, only there does the general capacity to possess a right become distinguished from concrete legal claims. Only the continual reshuffling of values in the market creates the idea of a fixed bearer of such rights. In the market, the person imposing liabilities simultaneously becomes himself liable. He changes roles instantaneously from claimant to debtor. Thus it is possible to abstract from the concrete differences between legal subjects and to accommodate them within one generic concept.[21]

Just as chance acts of exchange and primitive forms of exchange, such as the exchange of gifts, preceded the exchange transactions of developed commodity production, so too the armed individual, (or, more often, group of people, a family group, a clan, a tribe, capable of defending their conditions of existence in armed struggle), is the morphological precursor of the legal subject with his sphere of legal power extending around him. This close morphological link establishes a clear connection between the lawcourt and the duel, between the parties to a lawsuit and the combatants in an armed conflict. But as socially regulative forces become more powerful, so the subject loses material tangibility. His personal energy is supplanted by the power of social, that is, of class organisation, whose highest form of expression is the state.[22]

[21] This did not occur in Germany until the time when Roman law was adopted, which is, moreover, proven by the fact that there are no German words for the concepts 'person' and 'legal subject'. (Cf. Otto Friedrich von Gierke, *Das deutsche Genossenschaftsrecht*, 4 vols., Berlin, 1873; vol. II: *Geschichte des deutschen Körperschaftsbegriffs*, p. 30.)

[22] From this moment on, the figure of the legal subject begins to appear as something different from what it really is, that is to say not as the reflection of a relation arising behind people's back, but rather as an artificial creation of the human intellect. Yet the relations themselves become so habitual that they appear as indispensable conditions for every community. The idea that the legal subject is a purely artificial construct is as much a step in the direction of a scientific theory of law as the idea of the artificiality of money would be for economics.

In this form, the impersonal abstraction of state power functioning with ideal stability and continuity in time and space is the equivalent of the impersonal, abstract subject.

This power in the abstract has a perfectly real basis in the organisation of the bureaucratic machine, the standing army, the treasury, the means of communication, and so on. All of this presupposes the appropriate level of development of the productive forces.

Yet before calling on the machinery of the state, the subject depends on the stability of organically-based relationships. Just as the regular repetition of the act of exchange constitutes value as a universal category beyond subjective appraisal and arbitrary exchange ratios, so too the regular repetition of the same relations — custom — lends new significance to the sphere of subjective dominance by providing a basis for its existence in the form of an external norm.

Custom or tradition, as the supra-individual basis for legal claims, corresponds to the restrictive nature and inertia of the feudal social structure. Tradition, or custom, is by nature something confined to a particular, fairly narrow geographical area. Consequently, all rights were considered as appertaining exclusively to a given concrete subject or limited group of subjects. Marx says that in the feudal world, every right was a privilege. Every town, every estate, every guild lived according to its own law, which pursued the person wherever he went. This epoch completely lacked any notion of a formal legal status common to all citizens, to all men. This situation had its parallel in the economic sphere, in the self-supporting closed economies, the bans on import and export, and so on.

> Personality never had the same content universally. Originally rank, property, occupation, religious denomination, age, sex, physical strength, and so on generated such extensive inequality of legal rights that people could not see past the concrete differences to the constant elements of personality.[23]

Equality between subjects was assumed only for relations which were confined to a particular narrow sphere. Thus the members of one and the same estate were equal in the realm

[23] Gierke, *Genossenschaftsrecht,* op. cit., vol. II, p. 35.

of the law of that estate, members of one guild were equal in the realm of guild law. At this stage, the legal subject as the universal abstract bearer of every conceivable legal claim is in evidence only as a bearer of concrete privileges.

> Basically, however, even today the Roman dictum that the personality is inherently equal and that inequality is merely the outcome of an exceptional precept of positive law, has permeated neither legal life nor legal consciousness.[24]

Since there was no abstract concept of the legal subject in the Middle Ages, the concept of the objective norm, applicable to a wide, indeterminate circle of people, was also connected with the establishment of concrete privileges and freedoms.

As late as the thirteenth century we find no trace of any clear conception of the difference between objective law and subjective rights or legal powers (Berechtigungen). Wherever one looks, one finds that these two concepts are confused in the patents and charters given the towns by emperors and princes. The usual procedure for establishing any sort of universal regulation or norm was to furnish a particular part of the country or a section of the population with certain kinds of legal qualities. The well-known formula: 'Stadtluft macht frei' ('City air is liberating') has this character too. Judicial duels were abolished in the same manner. Town dwellers' rights to the usufruct of princely or imperial forests are a further example of rights conferred in this manner and regarded as being of an identical nature.

The same mixture of subjective and objective elements is evident in early municipal law. The municipal statutes were in part general charters, and in part an enumeration of isolated rights or privileges belonging to particular groups of citizens.

Only when bourgeois relations are fully developed does law become abstract in character. Every person becomes man in the abstract, all labour becomes socially useful labour in the

24 Ibid., p. 34. [This quotation from Gierke in the German edition replaces the following in the 3rd Russian edition, although the same page reference is given: 'Legal consciousness sees, at this stage, that the same or analogous rights are appropriated by individual personalities or collectives, but it does not produce the inference that therefore these personalities and collectives are one and the same in their attributes of rights.'—Ed.]

abstract,[25] every subject becomes an abstract legal subject. At the same time, the norm takes on the logically perfected form of abstract universal law.

The legal subject is thus an abstract owner of commodities raised to the heavens. His will in the legal sense has its real basis in the desire to alienate through acquisition and to profit through alienating. For this desire to be fulfilled, it is absolutely essential that the wishes of commodity owners meet each other halfway. This relationship is expressed in legal terms as a contract or an agreement concluded between autonomous wills. Hence the contract is a concept central to law. To put it in a more high-flown way: the contract is an integral part of the idea of law. In the logical system of juridical concepts, the contract is merely a form of legal transaction in the abstract, that is, merely one of the will's concrete means of expression which enable the subject to affect the legal sphere surrounding him. Historically speaking, and in real terms, the concept of the legal transaction arose in quite the opposite way, namely from the contract. Outside of the contract, the concepts of the subject and of will only exist, in the legal sense, as lifeless abstractions. These concepts first come to life in the contract. At the same time, the legal form too, in its purest and simplest form, acquires a material basis in the act of exchange. Consequently the act of exchange concentrates, as in a focal point, the elements most crucial both to political economy and to law. In exchange, Marx says, 'the content of this juridical relation (or relation of two wills) is itself determined by the economic relation.'[26] Once arisen, the idea of the contract strives to attain universal validity. The owners of commodities were of course proprietors even before they acknowledged one another as such, but in a different, organic, non-legal sense. 'Mutual recognition' is nothing more than an attempt to rationalise,

---

[25] 'For a society of commodity producers, whose general social relation of production consists in the fact that they treat their products as commodities, hence as values, and in this material (*sachlich*) form bring their individual, private labours into relation with each other as homogeneous human labour, Christianity with its religious cult of man in the abstract, more particularly in its bourgeois development, i.e. in Protestantism, Deism, etc., is the most fitting form of religion'. (Marx, *Capital*, vol. I, ed. cit., p. 172.)
[26] Ibid., p. 178. [Ed.]

with the aid of the abstract formula of the contract, the organic forms of appropriation based on labour, occupation and so on, which the society of commodity producers finds in existence at its inception. Considered in the abstract, the relationship of a person to a thing is totally devoid of legal significance. The jurists sense this when they try to construe the institution of private property as a relationship between subjects, in other words, between people. Yet they conceive of this relationship in a purely formal and, moreover, in a negative way, as a universal prohibition, which excludes everybody but the owner from using and disposing of the object.[27] This interpretation may be adequate for the practical purposes of dogmatic jurisprudence, but it is quite useless for theoretical analysis. In these abstract prohibitions, the concept of property loses any living meaning and renounces its own pre-juridical history.

If, then, development began from appropriation, as the organic, 'natural' relationship between people and things, this relation was transformed into a legal one as a result of needs created by the circulation of goods, primarily, that is, by buying and selling. Hauriou points out that, at first, trade by sea and by caravan did not create the need for property to be safeguarded. The distance separating the people involved in exchange from each other was the best protection against any claims. The establishment of permanent markets created the necessity for settling the question of right of disposal over commodities, and hence for property law.[28] The property title *mancipatio per aes et libram* in ancient Roman law shows that it arose simultaneously with the phenomenon of domestic ex-

---

[27] Thus, for example, Windscheid (*Lehrbuch des Pandektenrechts*, op. cit., vol. I, section 38), starting from the fact that law can exist only between persons, but not between a person and a thing, concludes that: 'The law of objects knows only vetoes . . . the content of the will power constituting this law, however, is a negative content: those opposed to the person with rights shall . . . refrain from affecting the object and shall not, by their attitude to the object, prevent that person from affecting the object.'

Siegmund Schlossmann (*Der Vertrag*, Leipzig, 1876) draws the logical conclusion from this view when he considers the concept of the law of objects as merely a 'terminological aid'. In contrast to this, Dernburg (*Pandekten*, op. cit., vol. I, section 22, note 5) rejects this view, according to which 'even property, apparently the most positive right of all', is supposed to have 'a purely negative content in legal terms'.

[28] Maurice Hauriou, *Principes de droit public*, p. 286.

change. Similarly, inheritance has only been established as a property title since the time when civil intercourse became interested in such a transfer.[29]

Writing of exchange, Marx says that one owner of commodities may appropriate another's commodity in exchange for his own only with the consent of the other commodity-owner.[30] This is exactly the notion which the representatives of the doctrine of natural law tried to express by attempting to give property a basis in the form of a primitive contract. They are right – not, of course, in the sense that such a contractual act did take place at some historical point in time, but in that natural or organic forms of appropriation assume a juridical 'rationale' in the reciprocal transactions of appropriation and alienation. In the act of alienation, abstract property right materialises as a reality. Any other employment of an object is related to some concrete form of its utilisation as a means of production or consumption. If, however, the object has a function as an exchange value, it becomes an impersonal object, a purely legal object, and the subject disposing of it becomes a purely legal subject. The contrast between feudal and bourgeois property can be explained by their different approach to circulation. Feudal property's chief failing in the eyes of the bourgeois world lies not in its origin (plunder, violence), but in its inertia, in the fact that it cannot form the object of a mutual guarantee by changing hands through alienation and acquisition. Feudal property, or property determined by estate, violates the fundamental principle of bourgeois society: 'the equal opportunity to attain inequality'. Hauriou, one of the most astute bourgeois jurists, quite rightly emphasises reciprocity as the most effective security for property, which can be brought about with the minimum use of external force. This mutuality, which is ensured by the laws of the market, lends property the quality of an 'eternal' institution. In contrast to this, the purely political security vouchsafed by the coercive machinery of state amounts to nothing more than the protection of specified personal stocks belonging to the owners – an aspect which has no fundamental significance. In the past, the class struggle has often resulted in a re-allocation of property, the expropriation

[29] Ibid., p. 287.
[30] Marx, *Capital*, vol. I, ed. cit., p. 178. [Ed.]

of usurers and large landowners.[31] Yet these upheavals, extremely unpleasant though they may have been for those groups and classes who were their victims, did not shake the foundations of private property, the economic framework linking economic units through exchange. The same people who had rebelled against property had no choice but to approve it next day when they met in the market as independent producers. That is the way of all non-proletarian revolutions. It is the logical consequence of the ideals of the anarchists. Whilst they do, of course, reject the external characteristic of bourgeois law — state coercion and the statutes — they preserve its inner essence, the free contract between autonomous producers.[32]

Thus, only the development of the market creates the possibility of — and the necessity for — transforming the person appropriating things by his labour (or by robbery) into a legal owner. There is no clearly defined borderline between these two phases. The 'natural' changes into the juridical imperceptibly, just as armed robbery blends quite directly with trade.

Karner's definition of property differs from this. According to him,

> property is *de jure* nothing but the power of disposal of a person A over an object N, the mere relation between individual and natural object which, according to the law, affects no other object and *no other person* (emphasis mine, E.P.). The object is private property, the individual a private person, and the law is private law. This was in accordance with the facts in the period of simple commodity production . . .[33]

[31] This gives rise to Engels' remark: 'It is thus entirely true that for 2,500 years private property could be protected only by violating property rights.' (Friedrich Engels, 'The Origin of the Family, Private Property and the State', in: Marx and Engels, *Selected Works*, vol. III, 1970, p. 281.)

[32] Thus Proudhon, for example, declares: 'I want the contract, but not laws; for me to be free, the entire social structure must be altered on the basis of reciprocal contract.' (Pierre Joseph Proudhon, *Idée générale de la Révolution au XIXe siècle: Choix d'études sur la pratique révolutionnaire et industrielle*, Paris, 1851, p. 138). Yet he is forced to add shortly after this: 'The norm according to which the contract shall be fulfilled will not be based exclusively on justice, but also on the common will of people living together. This will shall ensure the fulfilment of the contract, even by force if necessary.' (Ibid., p. 293.)

[33] Josef Karner (pseudonym of Karl Renner), *The Institutions of Private Law and their Social Functions*, edited by O. Kahn-Freund, trans-

This entire passage is a misconception from beginning to end. Karner is here re-creating that old favourite – the paradigm of Robinson Crusoe. Yet one wonders what point there is in two Robinsons, of whom one does not know the other exists, conceiving of their relationship to objects *in a legal fashion*, when it is an exclusively *factual relationship*. This law of the isolated individual merits comparison with the proverbial value of the 'glass of water in the desert'. Both value and property law are engendered by one and the same phenomenon: the circulation of products which have been transformed into commodities. Property in the legal sense did not arise because it occurred to people to invest one another with this legal capacity, but because they were able to exchange commodities only in the guise of property-owners. 'The unlimited power of disposal over objects' is nothing but the reflection of unlimited commodity circulation.

Karner states that the property-owner 'is quick-witted enough to cultivate the legal aspect of his right. He alienates the object'.[34] It does not occur to Karner that the 'juridical' only begins with this 'cultivation'; without it, appropriation does not transcend the limits of natural, organic appropriation.

Karner admits that 'sale and purchase, loan, deposit, rent existed previously, yet their range was very small, with regard to the *persona* as well as to the *res*'.[35] Indeed these various legal forms of the circulation of goods were in existence at such an early date that we have a precise formulation of lending and borrowing even before the formula for property itself had been elaborated. This fact alone is enough to provide the key to a correct understanding of the legal nature of property.

Yet Karner believes that people were property-owners even prior to buying and selling or pawning things. The relations we have mentioned appear to him merely as 'quite secondary, makeshift institutions, stop-gaps of petty bourgeois property'. In other words, he proceeds from the assumption of totally isolated individuals who have hit on the idea (one knows not why) of creating a 'common will', and – in the name of this

---

lated by A. Schwarzschild, London: Routledge & Kegan Paul Ltd., 1949, pp. 266-267.
[34] Ibid., p. 268.
[35] Ibid.

common will – of ordering everyone to refrain from assaults on objects belonging to someone else. Later – when they have realised that the property-owner cannot be regarded as entirely self-sufficient, either as a labourer or as a consumer – these isolated Robinsons decide to supplement property through the institutions of buying and selling, borrowing, lending and so forth. This purely rational scheme of things stands the actual development of things and concepts on its head.

Karner is here quite simply reproducing the so-called Hugo-Heysian system of interpretation of law. This too starts out in just the same way, from man subjugating the objects of the external world (law of objects), thence proceeding to the exchange of services (law of obligations), to arrive ultimately at the norms which regulate man's situation as member of a family and the fate of his assets after his death (family law and law of inheritance). Man's relationship to the things produced by him, or acquired by conquest, or forming, as it were, part of his personality (weapons, jewellery), has undoubtedly been a factor in the historical development of private property. This relationship represents property in its primitive, crude and limited form. Private property first becomes perfected and universal with the transition to commodity production, or more accurately, to capitalist commodity production. It becomes indifferent to the object and breaks off all ties with organic social groupings (family, tribe, community). It appears in its most universal sense as the 'external sphere of freedom',[36] that is, as the practical manifestation of the abstract capacity to be the subject of rights.

Property in this purely legal form has little logical connection with the organic natural principle of private appropriation resulting from personal expenditure of energy, or as the precondition for personal use and consumption. The fragmentation of the economic totality in the market renders the bond between the property-owner and his property just as abstract, formal, qualified and rationalistic, as the relationship of a person to the product of his labour (for example to the plot of land he cultivates himself) is elementary and accessible even to the most primitive turn of mind.[37] Whilst there is a direct

[36] Hegel's *Philosophy of Rights*, ed. cit., section 41ff. [Ed.]
[37] It is for precisely this reason that the apologists of private property

morphological connection between these two institutions: private appropriation as the precondition for unlimited personal usage, and private appropriation as the condition for subsequent alienation in the act of exchange, nonetheless, logically speaking they are two different categories, and the word 'property' used to describe both, creates more confusion than clarity. Capitalist landed property, for example, does not presuppose any kind of organic bond between the land and its owner. On the contrary, it is only conceivable when land changes hands with complete freedom.

The concept of landed property itself emerges simultaneously with individually alienable landed property. The common land of the *Allmende* was originally in no way the property of a legal person – such a concept did not even exist – but was for the usage of Mark-community members as a collective.[38]

Capitalist property is basically the freedom to transform capital from one form to another, the transfer of capital from one sphere to another for the purpose of gaining the highest possible unearned income. This freedom of disposition inherent in capitalist property is inconceivable without the existence of propertyless individuals, in other words, of proletarians. The legal form of property is not at all incompatible with the fact of the expropriation of a large number of citizens, for the capacity to be a legal subject is a purely formal capacity. It qualifies all people as being equally 'eligible for property', but in no way makes property-owners of them. Marx's *Capital* illustrates this dialectic of capitalist property brilliantly, both when it is absorbed by 'fixed' legal forms, and when it explodes these forms by the direct use of violence (in the period of primitive accumulation). Karner's analysis, which we have been discussing, has very little to offer in this respect which is new as compared with volume one of *Capital*. Where Karner does try to be original, he only creates confusion. We have already drawn attention to his attempt to abstract property from that aspect which constitutes it as juridical, that is, from exchange. This purely formal conception carries yet another misconcep-

---

are particularly fond of appealing to this primitive relation, for they know that its ideological force far outweighs its economic significance for modern society.

[38] Cf. Gierke, *Genossenschaftsrecht*, ed. cit., vol. II, p. 146.

tion in its wake: in analysing the transition from petty bourgeois to capitalist property, Karner declares:

> The legal institution of property . . . has undergone an extensive development in a relatively short period. It has suffered a drastic transformation which has not, however, been accompanied by noticeable modifications of its legal structure.

and directly after this concludes:

> The legal institution remains the same, as regards its normative content, but it no longer retains its former social functions.[39]

One wonders which institutions Karner means. If he is referring to the abstract formula of Roman law, then of course there is nothing in that which could be changed. But this formula has governed petty property only in the epoch of developed bourgeois capitalist relations.

However, if we consider guild trade and the peasant economy in the epoch of serfdom, then we shall find quite a number of norms limiting property right. A possible counter-argument is, of course, that these limitations are all of a public-law nature and do not affect the institution of property as such. But in that case the entire argument is reduced to the assertion that a particular abstract formula is identical to itself. Nonetheless, even feudal and guild — in other words, limited — forms of property revealed their function to be one of absorbing other people's unpaid labour. The property arising from simple commodity production, with which Karner contrasts the capitalist form of property, is an abstraction as bald as simple commodity production itself. For the transformation of even a portion of products into commodities, and the emergence of money, together create the conditions for the appearance of usurer's capital which, as Marx puts it,

> belongs together with its twin brother, merchant's capital, to the antediluvian forms of capital, which long precede the capitalist mode of production and are to be found in the most diverse economic formations of society.[40]

[39] Karner, *Institutions of Private Law*, ed. cit., pp. 252 and 257.
[40] Marx, *Capital*, vol. III, ed. cit., p. 580.

Hence we can reach a conclusion diametrically opposed to Karner's, namely: norms vary, but their social function remains unchanged.

As the capitalist mode of production develops, the property-owner gradually rids himself of technical production functions, thereby losing absolute legal sway over capital. In a joint-stock company, the individual capitalist is merely the bearer of a title to a certain quota of unearned income. His economic activity as a proprietor is almost totally limited to the sphere of unproductive consumption. The main bulk of the capital becomes an utterly impersonal class force. To the extent that this mass of capital participates in market transactions – which presupposes that its individual constituent parts are autonomous – these autonomous components appear as the property of legal persons. In reality, the whole bulk of the capital is controlled by a relatively small group of the largest capitalists, who act, moreover, not in person, but through their paid representatives or authorised agents. At this point, the juridically distinct form of property no longer reflects the real state of affairs, since, by means of share participation and control and so forth, actual dominance extends far beyond the purely legal framework. Here we come close to that moment when capitalist society is ready to turn into its opposite, the indispensable precondition for which is the class revolution of the proletariat.

Long before this revolution, however, the development of the capitalist mode of production based on the principle of free competition results in this latter principle being turned into its opposite. Monopolistic capitalism creates the preconditions for an entirely different economic system, in which the momentum of social production and reproduction is effected, not by means of individual transactions between autonomous economic units, but with the help of a centralised, planned organisation. This organisation is brought into being by trusts, combines, and other monopolistic associations. The Great War witnessed an embodiment of these tendencies when private capitalist and state organisations interlocked to form a powerful system of bourgeois state capital. This practical modification of the legal fabric could not leave theory untouched. In the rosy dawn of its evolution, industrial capitalism surrounded the principle of legal subjectivity with a halo by elevating it to the level of an absolute attribute of the human personality. Nowadays people are beginning to regard this principle rather as a purely technical determinant, which is well-suited to 'distinguishing risks and liabilities' or, alternatively, they pose it simply as a speculative hypothesis lacking any material basis. Since this latter approach directed its fire at legal individualism, it won the sympathies of various Marxists, who were of the opinion that it contained the elements of a new, 'social' legal theory corresponding to the interests of the proletariat. Obviously such an evaluation demonstrates a purely formal attitude to the problem. In any case, the theories mentioned do not provide any criteria whatever for a genuine sociological interpretation of the individualistic categories of bourgeois law, which they criticise, not from the point of view of the proletarian conception of socialism, but from the standpoint of the dictatorship of finance capital. The social significance of these doctrines is that they justify the modern imperialist state and its methods, particularly those employed in the last War. It should therefore come as no surprise to us that an American jurist draws similar 'socialist'-sounding conclusions precisely on the strength of the lessons of the World War, that most reactionary and rapacious of wars in recent history.

> The individual's rights to life, freedom and property have no absolute or abstract existence; they are rights which exist, from the legal standpoint, only because the state protects them, and which are, as a result, entirely subject to the authority of the state.[41]

Seizure of political power by the proletariat is the fundamental prerequisite of socialism. Nevertheless, experience has shown that planned production and distribution cannot replace market exchange and the market as the link between individual economic units overnight. Were this possible, then the legal form of property would be historically absolutely done for. It would have completed the cycle of its development and returned to its point of origin, to objects of direct, individual use; that is, it would in practice once more have become a primitive rela-

41 E. A. Harriman, 'Enemy Property in America', in *The American Journal of International Law*, 1924, vol. I, p. 202.

tion. And as a consequence of this the legal form as such would also be condemned to death.[42] So long as the task of building a unified planned economy has not been completed, so long as the market-dominated relationship between individual enterprises and groups of enterprises remains in existence, the legal form too will remain in force. It is hardly necessary to make specific mention of the fact that the form of private property which corresponds to the means of production in the economy of small farmers and craftsmen remains almost entirely unchanged during the transition period. But even in nationalised large-scale industry, application of what is called the principle of 'economic calculation' means the formation of autonomous units whose relationship to other economic units is mediated through the market.

To the extent that state enterprises are subject to the conditions of circulation, transactions between them take on the form, not of technical co-ordination, but of legal transactions. As a result, the purely legal, or juridical, regulation of relations becomes both possible and necessary. In conjunction with this, direct, or administrative, technical management is likewise preserved, and is undoubtedly strengthened over time through being subjected to a general plan of the economy.

On the one hand, therefore, we have economic life functioning in terms of the categories of natural economy, with a social link between units of production which appears in a rational, undisguised form (that is to say, not in commodity form). The method corresponding to this involves direct, or technically-determining prescriptions in the form of programmes, plans for production and distribution, and so forth. Such prescriptions are concrete, and are continually being modified in accordance with changing conditions. On the other hand, we have a relationship between economic units expressed in the form of the value of commodities in circulation and consequently in the form of legal transactions. Corresponding to this relationship, we have the creation in turn of more or less fixed and unchanging formal limitations on, and regulations for, legal inter-

[42] The subsequent process of transcending the legal form would be reduced to a gradual transition from equivalent distribution (a specific quantity of social products for a given quantity of labour) to the formula of developed communism: 'from each according to his abilities, to each according to his needs'.

course between autonomous subjects (civil code, perhaps also a commercial code), and of organs which help to sort out tangles in such transactions by means of judgments in lawsuits (courts, arbitration committees, and so on). Obviously the first of these tendencies offers no long-term prospects for the legal profession. The victory, by degrees, of this tendency means the gradual withering away of the legal form altogether. One can, of course, argue that a production programme, for example, is also a public-law norm, since it emanates from the state authority, has binding force, creates rights and obligations, and so on. It is true that, so long as the new society comprises elements of the old, that is, of people who conceive of the social relation purely as a means to their private ends, even simple, rational, technical instructions will necessarily assume the form of a supra-individual, external force. To quote Marx, political man will still be 'abstract, artificial man'. The more radically relations based on commodity exchange and the huckstering mentality have been overcome (in the realm of production), the sooner the day of final liberation will come, about which Marx said, in his article 'On the Jewish Question':

> Only when the real, individual man re-absorbs in himself the abstract citizen, and as an individual human being has become a *species-being* in his everyday life, in his particular work, and in his particular situation, only when man has recognised and organised his '*forces propres*' as *social forces*, and consequently no longer separates social power from himself in the shape of *political* power, only then will human emancipation have been accomplished.[43]

These are the perspectives for the unknown future. In dealing with our transition period, we must draw attention to the following. Whilst in the epoch of dominance by impersonal finance capital conflicts of interest continue to exist between individual groups of capitalists (who have their own and other people's capital at their disposal), under proletarian dictatorship, contrary to this, conflicts of interest are abolished within nationalised industry, despite the continuance of market exchange. The distinction between, or autonomy of, individual

---

[43] Marx, 'On the Jewish Question', in Marx and Engels, *Collected Works*, vol. III, 1975, pp. 167-168.

economic organisms (on the model of the autonomy of private production) is *retained as a method only.* In this way, those *quasi* private-enterprise relations which arise between state industry and small economic units, as well as amongst individual enterprises and groups of enterprises within state industry itself, are confined within strict limits, determined at any given moment by the successes achieved in the sphere of the planned direction of the economy. Hence, in our transition period, the legal form as such does not contain within itself those unlimited possibilities which lay before it at the birth of bourgeois-capitalist society. On the contrary, the legal form only encompasses us within its narrow horizon for the time being. It exists for the sole purpose of being utterly spent.

The task of Marxist theory consists of verifying this general conclusion and of following up the concrete historical material. Development cannot proceed evenly in all areas of social life. That is why painstaking labour in observation, comparison and analysis is absolutely indispensable. Only when we have closely examined the tempo and form of transcending value relations in the economy and, simultaneously, of the withering away of private-law aspects of the legal superstructure and, finally, the progressive dissolution of the legal superstructure itself, conditioned by these fundamental processes, only then shall we be able to say that we have clarified at least one aspect of the process of building the classless culture of the future.

5. *Law and the State*

Legal intercourse does not 'naturally' presuppose a state of peace, just as trade does not, in the first instance, preclude armed robbery, but goes hand in hand with it. Law and self-help, those seemingly contradictory concepts are, in reality, extremely closely linked. This is true, not only of the most ancient epoch of Roman law, but also of later periods. Modern international law includes a very considerable degree of self-help (retaliatory measures, reprisals, war and so on). Even in the 'well-ordered' bourgeois state, rights are secured – in the opinion of an astute jurist like Hauriou – by every citizen at his own risk. Marx formulates this even more succinctly in his 'Introduction to the Critique of Political Economy': 'Even club-law is law'.[1] This is not a paradox, for law, like exchange, is an expedient resorted to by isolated social elements in their inter-course with one another. At different stages cf history this iso-lation can be more or less pronounced, but it can never vanish altogether. Thus the enterprises belonging to the Soviet state, for example, actually fulfil a communal task; yet, because in their work they are forced to adhere to market methods, each one of them has separate interests. They confront one another as buyer and seller, do business at their own risk, and, as a result, must inevitably engage in *legal intercourse* with one another. The ultimate victory of planned economy will trans-form their relationship into an exclusively technical expedient,

[1] Karl Marx, *Grundrisse: Introduction to the Critique of Political Economy*, 1973 ed., p. 88: 'The principle of might makes right . . . is also a legal relation'.

thereby doing away with their 'legal personality'. Therefore, whenever people portray legal intercourse as organised and well-ordered, and thus equate law with legal order, they forget, in so doing, that this order is actually a mere tendency and end result (by no means perfected at that), but never the point of departure and prerequisite of legal intercourse. The very state of peace, which to abstract legal thought seems homogeneous and undifferentiated, was not so at all in the early stages of legal development. Ancient Germanic law recognised varying degrees of peace: peace in the household, peace within the enclosure, peace within the boundaries of the settlement, and so on. The greater or lesser degree of pacification was expressed in the varying severity of the punishment meted out to the person guilty of breach of the peace.

The state of peace becomes a necessity when exchange becomes a regular phenomenon. In cases where there were insufficient preconditions for keeping the peace, the persons engaged in exchange preferred to each inspect the commodities separately, in the absence of the other party, rather than meeting together personally. In general, however, trade requires that not only the commodities, but the people too come together. In the epoch of the gens-system, every stranger was regarded as an enemy: he was free game, just like the beasts of the forest. Only the custom of hospitality provided an opportunity for intercourse with alien tribes. In feudal Europe, the Church attempted to check the incessant private wars by proclaiming the so-called *treuga dei* at certain intervals.[2] Simultaneously, markets and trading centres began to be endowed with appropriate special concessions. Merchants going to market received safe conduct, their property was protected against arbitrary seizure. At the same time, the fulfilment of contracts was guaranteed by special judges. In this way, a special *jus mercatorum* or *jus fori* arose, which formed the basis of later municipal law.

Originally, market places and fairs were part and parcel of

[2] It is characteristic of the church that by prescribing 'divine peace' for certain days, it thereby sanctioned private wars for the rest of the time. In the eleventh century, it was suggested that these wars be completely abolished. Gérard, bishop of Combres, protested vehemently against this, declaring that the demand for continuous divine peace was contrary to 'human nature'. (Cf. S. A. Kotlyarevsky, *Power and Law, Vlast' i pravo*, Moscow, 1915, p. 189.)

feudal demesnes, and were simply advantageous sources of income for the local feudal lord. Whenever a place was granted market peace,[3] the sole purpose of this was to fill the purse of some feudal lord, so that it was in his private interest. Yet thanks to its new role as guarantor of the peace indispensable to the exchange transaction, feudal authority took on a hue which had hitherto been alien to it: it went *public*. The feudal or patriarchal mode of authority does not distinguish between the private and the public. The feudal lord's public rights with regard to his serfs were simultaneously his rights as a private owner, whereas his private rights, on the contrary, may be interpreted, if one so wishes, as political, and therefore public rights. This is exactly the way in which many people (including Gumplowicz) interpret the *jus civile* of ancient Rome as a public law, since it was founded on, and originated in membership of a gens-organisation. In fact, we are faced here with an embryonic form of law, which has not yet developed within it the contradictory, yet correlate, determinants 'private law' and 'public law'. That is why any form of power which bears traces of patriarchal or feudal relations is also characterised by the predominance of theological rather than legal aspects. Only the development of trade, and of the money economy, make the juridical, or rationalistic, interpretation of the phenomenon of power possible. It is these economic forms which first introduce the contradiction between public and private life, a contradiction which assumes, over time, an 'eternal' and 'natural' character, and forms the basis of every juridical theory of power.

The 'modern' state (in the bourgeois sense) comes into being at that point in time when the organisation of power by groups or classes encompasses a sufficiently expanded activity in market transactions.[4] Thus, in Rome, trade with foreigners, resi-

---

[3] 'The rapid extension of the king's peace till it becomes, after the Norman Conquest, the normal and general safeguard of public order, seems peculiarly English.' . . . 'The churches have their peaces . . . the sheriff has his peace, the lord of a soken has his peace, may, every householder has his peace. . . . If the king can bestow his peace on a privileged person by his writ of protection, can he not put all men under his peace by proclamation?' (Pollock, F. and F. W. Maitland, *The History of English Law before the Time of Edward I*, 2nd ed., Cambridge: Cambridge University Press, 1911, vol. I, p. 45; vol. II, p. 454). [Ed.]

[4] Cf. Maurice Hauriou, *Principes de droit public*, p. 272.

dent aliens (*peregrini*) and others, demanded acknowledgement of legal capacity in the civil sphere in people who were not members of the gens-association. This already presupposes, however, a distinction between public and private law.

The earliest and most complete separation between the public-law principle of territorial sovereignty and private land ownership occurs in medieval Europe, within the city walls. It is there that the material and personal obligations pertaining to land disintegrate earlier than anywhere else into taxes and obligations in favour of the municipality on the one hand, and into rent based on private property on the other.[5]

Effective power acquires a markedly juridical, public character, as soon as relations arise in addition to and independently of it, in connection with the act of exchange, that is to say, private relations par excellence. By appearing as a guarantor, authority becomes social and public, an authority representing the impersonal interest of the system.[6]

The state as an organisation of class rule, and for waging external wars, neither needs nor admits of any legal interpretation. This is an area where so-called *raison d'état* holds sway, which is nothing but the principle of naked expediency. In contrast to this, power as a guarantor of market exchange not only employs the language of law, but also functions as law and law alone, that is, it becomes one with the abstract, objective norm.[7] Consequently every juridical theory of the state which attempts to encompass *all* state functions is nowadays inadequate. It cannot accurately reflect all the facts of state life, it gives a purely ideological, that is, a distorted reflection of reality.

[5] Cf. Otto Friedrich von Gierke, *Das deutsche Genossenschaftsrecht*, 4 vols., Berlin, 1873; vol. II: *Geschichte des deutschen Körperschaftsbegriffs*, p. 648.

[6] Although both the Western feudal lords and the Russian princes were actually quite oblivious of this elevated mission of theirs and saw their function as guardians of order merely as a source of income, subsequent bourgeois historians could not, of course, resist attributing fictional motives to them. After all, these historians themselves saw bourgeois aspirations and the resulting public nature of authority as an eternal and immutable norm.

[7] The objective norm is thereby interpreted as the universal conviction of those subject to the norm. Law is supposedly the universal belief of persons engaged in legal intercourse. The origin of a legal situation is therefore supposed to be the origin of a universal belief which has bind-

Class rule, in both its organised and its unorganised forms, is much more far-reaching than the sphere which can be designated as the state authority's official sphere of jurisdiction. The dominance of the bourgeoisie is expressed in the dependence of governments on banks and capitalist associations, and in the dependence of every individual worker on his employer, as well as in the fact that the personnel of the civil service is closely interlinked with the ruling class. All these facts – and there are any number of them – have no kind of official legal expression at all, yet in their consequences they coincide with the facts that do indeed find official legal expression in the subordination, for instance, of those very workers, to the laws of the bourgeois state, to the orders and decrees of its organs, to the sentences of its courts, and so on. Thus there arises, besides direct, unmediated class rule, indirect, reflected rule in the shape of official state power as a distinct authority, detached from society. This raises the problem of the state, which poses no lesser difficulties for analysis than the problem of the commodity.

In his *Origins of the Family, Private Property and the State*, Engels regards the state as the expression of the fact that society has become entangled in insoluble class antagonisms.

But in order that these antagonisms, classes with conflicting interests, might not consume themselves and society in fruitless struggle, it became necessary to have a power seemingly standing above society that would alleviate the conflict and keep it within the bounds of 'order'; and this power, arisen out of society but placing itself above it, and alienating itself more and more from it, is the state.[8]

---

ing force and is subject to being realised. (Georg Friedrich Puchta, *Vorlesungen über das heutige römische Recht*, edited by A. A. F. Rudorff, 6th ed., 2 vols., Leipzig, 1873.) This formula in its apparent universality is actually merely the ideal reflection of the conditions of market commerce. Without these, the formula is meaningless. Surely no one would have the nerve to assert that the legal position of the helots in Sparta, for example, results from their universal belief, which has acquired binding force. (Cf. Ludwik Gumplowicz, *Rechtsstaat und Sozialismus*, Innsbruck, 1881.)

[8] Friedrich Engels, 'The Origin of the Family, Private Property and the State', in: Marx and Engels, *Selected Works*, vol. III, 1970, p 327

A part of this argument is not quite clear, as will become apparent in what follows. Engels maintains that state power of necessity falls into the hands of the most powerful class, 'which, through the medium of the state, becomes also the politically dominant class'.[9] This proposition gives rise to the assumption that state power comes into being, not as a class force, but as something standing above classes, saving society from disintegration, and becoming the object of usurpation only after its emergence. Such a conception would, of course, fly in the face of historical facts. We know that the machinery of power has always been created by the ruling class. It is our opinion that Engels would himself have rejected such an interpretation of his words. But be that as it may, the formula posited by him still remains unclear. According to this formula, the state emerges because the classes would otherwise mutually destroy one another in bitter struggle, wrecking the whole of society in the process. It follows that the state emerges in a situation where neither of the two conflicting classes is in a position to force a decisive victory. This implies either that the state perpetuates the relationship of equilibrium and is therefore a force standing above the classes, which we cannot admit, or that it is the result of the victory of one class or another. However, were the latter true, society would have no further need for the state, since the decisive victory of one class would restore equilibrium, thus saving society. All these controversies mask one and the same fundamental question: why does class rule not remain what it is, the factual subjugation of one section of the population by the other? Why does it assume the form of official state rule, or – which is the same thing – why does the machinery of state coercion not come into being as the private machinery of the ruling class; why does it detach itself from the ruling class and take on the form of an impersonal apparatus of public power, separate from society?[10] It is

---

[9] Ibid, p. 328.

[10] In our times of heightened revolutionary struggle, we can observe how the official machinery of the bourgeois state apparatus retires into the background as compared with the volunteer corps of the fascists and others. This further substantiates the fact that, when the balance of society is upset, it seeks salvation not in the creation of a power standing above society, but in the maximal harnessing of all forces of the classes in conflict.

not enough to confine ourselves to pointing out that it is *advantageous* to the ruling class to erect an ideological smoke-screen, and to conceal its hegemony beneath the umbrella of the state. For although such an elucidation is undoubtedly correct, it still does not explain how such an ideology could arise, nor, therefore, does it explain why the ruling class has access to it. For the conscious exploitation of ideological forms is of course something separate from their emergence, which usually occurs independently of people's will. If we wish to expose the roots of some particular ideology, we must search out the material relations which it expresses. In the process we shall, moreover, encounter one of the fundamental differences between the theological and the juridical interpretation of the concept 'state power'. In the former interpretation we are dealing with fetishism of the first order; consequently we shall not succeed in discovering anything at all in the corresponding ideas and concepts other than an ideological reproduction of reality, in other words the same factual relations of dominance and subservience. In contrast to this conception, the legal view is one-dimensional; its abstractions are the expression of only one of the facets of the subject as it actually exists, that is, of commodity-producing society.

In his book *Problems of the Marxist Theory of Law*, Razumovsky accuses me of transposing questions of dominance and subservience to the ambiguous realm of the 'reproduction of reality', and of not granting them their rightful place in the analysis of the category of law.[11] It seems to me that after Feuerbach and Marx there is no longer any need to debate the fact that religious or theological thought represents a 'duplication of reality'. I do not see anything ambiguous in that. On the contrary, the facts of the matter are transparently clear: the serf's subjugation by a feudal lord was the direct and unmediated result of the feudal lord being a landowner with an armed force at his disposal. This direct dependence, this seigneurial relation, increasingly assumed an ideological veneer, with the feudal lord's power being increasingly deduced from a divine, supra-human authority: 'No power if not God-given'. The wage-labourer's subjection to, and dependence on,

11 Isaak Petrovich Razumovsky, *Problems of the Marxist Theory of Law (Problemy marksistskoy teorii prava)*, Moscow, 1925.

the capitalist has a similar immediacy: congealed, dead labour here dominates living labour. Yet this worker's subjugation by the capitalist state is not a mere ideological duplication of his dependence on the individual capitalist. It is not the same, firstly because we have here a special apparatus, separate from the representatives of the ruling class, which stands above every individual capitalist and functions as an impersonal force. Secondly, it is not the same because this impersonal force does not mediate every individual exploitative relation; the wage worker is not actually politically and legally *compelled* to work for a *particular* entrepreneur; rather, he formally sells that entrepreneur his labour power on the basis of a free contract. In so far as the exploitative relation exists formally as a relationship between two 'autonomous' and 'equal' owners of commodities, of whom one, the proletarian, sells his labour power, and the other, the capitalist, buys it, political class power can take on the form of public authority. As we have already said, the principle of competition prevailing in the bourgeois-capitalist world does not allow any possibility of linking political power to the individual economic enterprise (in the way in which this power was linked to landed property under feudalism).

> Free competition, the freedom of private property, 'equality' in the market and the monopolisation of the means of production by one class combine to produce a new form of state power: democracy, which enables that class to come to power collectively.[12]

It is quite true that 'equality' in the market creates a specific form of power, but these phenomena are connected in quite a different way than Podvolotsky thinks. To begin with, power can remain the private affair of capitalist organisations, even without being linked to the individual enterprise. The industrial associations with their fighting funds, their black-lists, their lockouts and their regiments of strike-breakers, are undoubtedly organs of power which coexist with public, that is, with state power. Moreover, control within the enterprise remains the

---

[12] I. Podvolotsky, *The Marxist Theory of Law (Marksistskaya teoriya prava)*, 1923, p. 33.

private affair of each individual capitalist. The establishment
of labour regulations is an act of private legislation; in other
words, it is a piece of pure feudalism. This remains true despite
the lengths to which bourgeois jurists go in order to tart it up
in a modern fashion by creating the fiction of the so-called
contrat d'adhésion, or of a special mandate which the capitalist
is supposedly granted by the organs of official power, so that he
may 'successfully perform the socially necessary and expedient
functions of the enterprise.'[13]

The analogy with feudal relations is, nevertheless, not absolutely apposite in the given example, for, as Marx says:

> The authority assumed by the capitalist as the personification
> of capital in the direct process of production, the social func-
> tion performed by him in his capacity as manager and ruler
> of production, is essentially different from the authority
> exercised on the basis of production by means of slaves, serfs,
> etc.
> . . . on the basis of capitalist production, the mass of direct
> producers is confronted by the social character of their
> production in the form of strictly regulating authority and a
> social mechanism of the labour-process organised as a com-
> plete hierarchy — this authority reaching its bearers, however,
> only as the personification of the conditions of labour in con-
> trast to labour, and not as political or theocratic rulers as
> under earlier modes of production . . .[14]

Thus it is possible for the relations of dominance and subservience to exist in the capitalist mode of production too,
without deviating from the concrete form in which these relations figure: as the domination of the producers by the conditions of production. Yet precisely the fact that they do not
appear in veiled form here — as they do in slavery and serfdom[15]
— makes them incomprehensible to the jurist.

13 Cf. Tal', 'The Legal Nature of the Organisation or of Labour Regulations in the Enterprise' (Yuridicheskaya priroda organizatsii ili vnutren-
nego poryadka predpriyatiya'), in: *Yuridichesky Vestnik*, 1915, no. 9.
14 Marx, *Capital*, vol. III, 1962 ed., p. 859.
15 Ibid., p. 810: '. . . where slavery or serfdom form the broad foundation of social production, as in antiquity and during the Middle Ages,
. . . the domination of the producers by the conditions of production is
concealed by the relations of dominion and servitude, which appear and

To the extent that society represents a market, the machinery of state is actually manifested as an impersonal collective will, as the rule of law, and so on. Every buyer and seller is, as we have already seen, a legal subject par excellence. The autonomous will of those engaged in exchange is an indispensable precondition wherever the categories of value and exchange value come into play. Exchange value ceases to be exchange value, the commodity ceases to be a commodity, if exchange ratios are determined by an authority situated outside of the internal laws of the market. Coercion as the imperative addressed by one person to another, and backed up by force, contradicts the fundamental precondition for dealings between the owners of commodities. In a society of commodity owners, and within the limits of the act of exchange, coercion is neither abstract nor impersonal -- hence it cannot figure as a social function. For in the society based on commodity production, subjection to one person, as a concrete individual, implies subjection to an arbitrary force, since it is the same thing, for this society, as the subjection of one owner of commodities to another. That is also why coercion cannot appear here in undisguised form as a simple act of expediency. It has to appear rather as coercion emanating from an abstract collective person, exercised not in the interest of the individual from whom it emanates -- for every person in commodity-producing society is egoistic -- but in the interest of all parties to legal transactions. The power of one person over another is brought to

---

are evident as the direct motive power of the process of production.'

Pahukanis expresses himself rather unclearly in this sentence. If we refer to *Capital* Vol. III, we find at the end of the chapter on the 'Trinity Formula' that Marx is concerned to point out that the masking of social relationships by forms such as capital is excluded in slavery and serfdom where it appears that the motive force of production arises from social relation of dominance and subservience.

What is obscured in the latter cases, however, is the domination of the producers by the conditions of production, whereas under capitalism the power of economic forces is evident, and furthermore, the relations of dominance and subservience flow *directly* from them. The authority of the capitalist is not politically or theoretically established but proceeds from his ownership of a factor of production. Presumably what Pashukanis is getting at is that authority coming forward as the rule of 'dead matter', only contingently related to its owner, does not have to be juridically privileged (everyone can own property if he can get it) and hence eludes jurists. [Ed.]

bear in reality as the force of law, that is to say as the force of an objective, impartial norm.

Bourgeois thought, which takes the framework of commodity-production for the eternal and natural framework of every society, therefore regards abstract state power as an appurtenance of any and every society.

This is most naïvely expressed by the theoreticians of natural law, who based their theory of power on the idea of the intercourse between autonomous and equal personalities and imagined that, in so doing, they were proceeding from the principles of social intercourse as such. In fact, all they have developed, in various keys, is the theme of a power representing the connecting link between autonomous owners of commodities. This explains the fundamentals of this doctrine, which were evident as early as Grotius. For the market, what really matters is that the commodity owners should engage in exchange; the power structure is something secondary and derived, externally imposed on the existing commodity owners. This is why the theoreticians of natural law do not regard state power as a phenomenon which arose historically and is therefore bound up with forces at work in the society in question, but think of it rather in an abstract and rationalistic manner. In intercourse between commodity owners, the necessity for authoritarian coercion arises whenever the peace is disturbed or a contract not fulfilled voluntarily. The doctrine of natural law therefore reduces the function of state power to keeping the peace, and sees the state's sole determining feature as being the instrument of law. After all, in the market, one owner of commodities is a commodity owner by the consent of the others, and all of them are commodity owners through their collective will. That is why the theory of natural law deduces the state from the contract between isolated individuals. This is the entire theory in skeleton form, which admits of the most diverse concrete variations, depending on the historical position, or the political sympathies and dialectical abilities of any one author. It allows of republican as well as monarchist deviations, and of altogether the most varying degrees of democratic and revolutionary '-isms'.

On the whole, though, this theory was the revolutionary banner under which the bourgeoisie fought out its revolutionary struggles against feudal society. That also decided the fate of the theory. Since the bourgeoisie became the ruling class, the revolutionary past of natural law has begun to arouse apprehension, and the prevailing theories cannot wait to consign it to oblivion. Of course the theory of natural law does not stand up to any kind of historical or socialist critique, for the picture it paints in no way corresponds to reality. Yet the odd thing is that the juridical theory of the state which superseded the natural law theory of the state, and which dispensed with the doctrine of the inherent and inviolable rights of people and citizens, thus acquiring the label 'positive', distorts actual reality just as much as its predecessor.[16] It is forced into this distortion, because every *juridical* theory of the state must of necessity posit the state as an independent power separated from society. It is this which constitutes the *juridical* nature of the theory.

Although in fact the activity of state organisations goes on in the shape of decrees and edicts emanating from individuals, juridical theory assumes firstly, that the state, not individuals, issues these orders, and secondly, that these orders are subordinate to the general norms of the code which, in turn, expresses the will of the state.[17]

On this point, the theory of natural law is not one whit more fanciful than any of the juridical theories of the state, even the

[16] It is not necessary for me to substantiate this argument explicitly, since I can refer the reader to Gumplowicz's critique of the juridicial theories of Labande, Jellinek and others (cf. Gumplowicz, *Rechtsstaat und Sozialismus*, op. cit. and *Geschichte der Staatstheorien*, Innsbruck, 1905), and further, to Vladimir Viktorovich Adoratsky's excellent work *On the State* (O gosudarstve), Moscow, 1923.

[17] Here we must point out a small contradiction. If not people, but the state itself acts, then why still place particular emphasis on subjection to the norms of that same state? . . . This is a repetition of one and the same thing. The theory of the organs of state is altogether one of the greatest stumbling blocks of legal theory. After the jurist has with great difficulty come to terms with the definition of the state, and endeavours to sail on without hindrance, a new snag awaits him – the concept of the 'organ'. Thus, in Jellinek for example, the state has no will, yet the organs of state do. One cannot but ask how, then, these organs arose? There is no state without organs. The attempt to evade the problem by conceiving of the state as a legal relation merely replaces the general problem with a series of special cases into which the problem disintegrates. For every concrete public law relation contains within it the same element of mystification which we find in the general concept of the 'state as a person'.

most positive of them, for the substantive element in the doctrine of natural law was that it posited, in addition to the various forms of mutual personal dependence (from which this doctrine abstracts), another form of dependence – on the impersonal collective will of the state.

Yet just this construct also forms the basis of the legal theory of the 'state as a person'. The element of natural law in the juridical theory of the state lies much deeper than was apparent to the critics of the doctrine of natural law. It lies in the very concept of *public* authority, that is, of an authority which belongs to no one in particular, stands above *everyone*, and addresses itself to *everyone*. In using this concept to orient itself, juridical theory inevitably loses touch with actual reality. The difference between the doctrine of natural law and most recent legal positivism consists only in the fact that the former is much more clearly aware of the logical connection between abstract state power and the subject in the abstract. It grasped the necessary connection between these mystically veiled relations of commodity-producing society and hence provided an example of classical clarity of construction. In contrast to this, would-be legal positivism is undecided even about its own logical premises.

The constitutional state (*Rechtsstaat*) is a mirage, but one which suits the bourgeoisie very well, for it replaces withered religious ideology and conceals the fact of the bourgeoisie's hegemony from the eyes of the masses. The ideology of the constitutional state is even more convenient than religious ideology, because, while it does not entirely reflect objective reality, it is still based on this reality. Power as the 'collective will', as the 'rule of law', is realised in bourgeois society to the extent that this society represents a market.[18] From this point of view, even police regulations can figure as the embodiment of the Kantian idea of freedom limited by the freedom of others.

18 It is well-known that Lorenz Stein contrasted the ideal state, standing above society, with the state absorbed by society, that is, to use our terminology, the class state. He characterised the feudal-absolutist state, which protects the privileges of landed property, and the capitalist state, which protects the privileges of the bourgeoisie, in the same way. But after deducting these historical realities, all that remains is the state as the fantasy of a Prussian official, or as the abstract guarantee of the conditions of exchange according to value. In historical reality, however, the 'constitutional state', that is to say the state standing above society,

The free and equal owners of commodities who meet in the market are free and equal only in the abstract relation of appropriation and alienation. In real life, they are bound by various ties of mutual dependence. Examples of this are the retailer and the wholesaler, the peasant and the landowner, the ruined debtor and his creditor, the proletarian and the capitalist. All these innumerable relationships of actual dependence form the real basis of state structure, whereas for the juridical theory of the state it is as if they did not exist. Further, the life of the state consists of the struggle between various political forces, between classes, parties, all manner of groupings; it is here that the real mainsprings of the machinery of state lie hidden. To juridical theory they are just as incomprehensible as the relationships mentioned above. The jurist may well be able to show a greater or lesser flexibility and ability to adapt to the facts; he can, for example, also take into account, besides statute law, the unwritten rules which have gradually arisen in the course of state practice, but that does not alter anything in principle in his fundamental approach to reality. A certain discrepancy between legal truth and the truth to which historical and sociological research aspires is unavoidable. This is due not only to the fact that the dynamic of social life overturns rigidified legal forms and that, as a result, the jurist is condemned always to complete his analysis far too late; even if he does remain up to date with the facts in his assertions, he renders these facts differently than the sociologist. For, so long as he remains a jurist, he starts from the concept of the state as an autonomous force, set apart from all other individual and social forces. From the historical and political point of view, the resolutions of an influential class or party organisation have a significance as great, and sometimes greater, than the decisions of parliament or of any other state organisation. From the legal point of view, facts of the first kind are, as it were, non-existent. In contrast to this, one can, by ignoring the legal standpoint, see in every parliamentary resolution not an act of state, but a decision reached by a particular group or clique

is only realised as its own opposite, as a 'committee for managing the common affairs of the whole bourgeoisie'. [Marx and Engels, 'Manifesto of the Communist Party', in Marx and Engels, *Selected Works*, vol. I, 1969, pp. 110-111. Ed.]

(acting in accordance with individual-egoistic or class-orientated motives, like every other collective). The extreme normativist Kelsen concludes from this that the state exists only in theory, as a closed system of norms or obligations. Such immateriality in the object of the theory of constitutional law must indeed alarm the practical jurists. For they sense, if not with their reason, then with their instinct, the undoubted practical validity of their concepts, precisely in this wicked world and not in the sphere of pure logic alone. The 'state' of the jurists is linked, despite its 'ideological nature', to an objective reality, just as the most fantastic dream is still based on reality.

This reality is primarily the machinery of state itself, with all its material and personal elements.

Before creating internally consistent theories, the bourgeoisie first constructed its state in practice. This process was inaugurated in the municipalities of Western Europe.[19] Whilst the feudal world made no distinction between the personal resources of the feudal lord and the resources of the political organisation, in the cities there arose, at first sporadically, and later as a permanent institution, the communal city purse;[20] the spirit of 'statedom' is given, so to speak, its material basis.

The establishment of state financial resources encouraged the appearance of people who live from these means: officials and functionaries. In the feudal epoch, administrative and judicial functions were performed by servants of the feudal lord. Public offices in the true sense of the word first appear in the municipalities; they are the material embodiment of the public nature of power. Authorisation in the private-law sense of a mandates to conclude legal transactions becomes separated from public office.

Absolute monarchy needed only to usurp this public form of power which had originated in the cities and to apply it to a wider area. Every additional improvement in bourgeois state-

---

[19] Cf. Kotlyarevsky, *Power and Law*, ed. cit., p. 193.
[20] The ancient Germanic Mark-community was not a legal person with property at its disposal. The public nature of the *Allmende* was expressed in the fact that it was used by all members of the Mark community. Collections for public needs were only made occasionally, and then strictly in accordance with need. If there was any surplus, it was used for communal entertainment. This custom shows how alien the idea of permanent public funds was.

dom, whether achieved by revolutionary outbreaks or by peaceful adaptation to feudal-monarchist elements, can be traced back to a single principle, according to which neither of two people exchanging in the market can regulate the exchange relation unilaterally; rather this requires a third party who personifies the reciprocal guarantees which the owners of commodities mutually agree to as proprietors, and hence promulgates the regulations governing transactions between commodity owners.

The bourgeoisie based its theories on this juridical concept of the state, which it attempted to put into practice. In the process of this realisation, however, it allowed itself to be guided by the notorious principle: * 'just as it has been done in the past, so it shall be done in the future'.[21]

For the bourgeoisie has never, in favour of purity of theory, lost sight of the fact that class society is not only a market where autonomous owners of commodities meet, but is at the

[21] The English bourgeoisie, which gained dominance in the world market earlier than anyone else, and which felt invulnerable as a result of its insular situation, was able to go further than anyone else in realising the 'constitutional state'. The most consistent realisation of the legal principle in the dialectical relationship between state power and individual subject, and the most effective guarantee that those in power cannot step out of their role as the personification of an objective norm, is the subjection of the organs of state to the jurisdiction of an independent court (not, of course, independent of the bourgeoisie). The Anglo-Saxon system is a kind of apotheosis of bourgeois democracy. Yet under different historical circumstances the bourgeoisie is, as it were, at worst also prepared to make do with a system which may be called the 'separation of property from the state', or 'Caesar-ism'. In this case, the ruling clique, with its unlimited, despotic caprice (which follows two directions, internally against the proletariat and externally in the form of an imperialist foreign policy), apparently creates a basis for the 'free self-determination of the personality' in bourgeois intercourse. Thus, in Kotlyarevsky's view for example, 'private law individualism' coexists with 'political despotism; the *code civil* originates in an epoch which is characterised not only by a lack of political freedom in the political system of France, but also by a certain indifference towards this freedom, which became vividly evident as early as the 18th Brumaire. Yet such private law freedom not only initiates a coming to terms with every aspect of state activity, but also lends the latter a certain character of legality'. (Kotlyarevsky, *Power and Law*, op. cit., p. 171). For a brilliant characterisation of Napoleon's relationship to bourgeois society, see Marx, 'The Holy Family', in: Marx and Engels, *Collected Works*, vol. IV, 1975, p. 123.

*[We render this 'principle' from the 3rd Russian edition; the German replaces it with 'insofar as' (*insofern als*); Ed.]

same time the battlefield of a bitter class war, where the machinery of state represents a very powerful weapon. On this battlefield, relations do not appear to be in the least in the spirit of Kant's definition of law as a minimal limitation of the freedom of the personality indispensable to human coexistence.

Gumplowicz is right here when he claims that this kind of law never existed, for

the measure of some people's 'freedom' is solely dependent on the measure of their domination by others. The norm is determined, not by the possibility of coexistence, but by the domination of some by others.

The state as a power factor in internal and foreign policy – that is the correction which the bourgeoisie was forced to make to the theory and practice of its 'constitutional state'. The more the hegemony of the bourgeoisie was shattered, the more compromising these corrections became, the more quickly the 'constitutional state' was transformed into a disembodied shadow, until finally the extraordinary sharpening of the class struggle forced the bourgeoisie to discard the mask of the constitutional state altogether, revealing the nature of state power as the organised power of one class over the other.

6. Law and Morality

For the products of human labour to be able to relate to each other as values, it is necessary for people to relate to each other as autonomous and equal personalities.

If one person is in another's power, that is, if he is a slave, his labour ceases to create, and form the substance of, values. The labour power of slaves, like the labour power of domestic animals, transfers only a certain portion of the costs of its own production and reproduction to the product.

Tugan-Baranovsky draws the inference from this that one can comprehend political economy only by starting out from the central ethical concept of the supreme value and thus identical worth of the human personality.[1] As we know, Marx infers the opposite: he relates the ethical idea of the equal worth of human personalities to the commodity form, in other words, he derives this idea from the practical equalisation of all forms of human labour.

Man as a moral subject, that is as a personality of equal worth, is indeed no more than a necessary condition for exchange according to the law of value. Man as a legal subject, or as a property-owner, is a further necessary condition. Finally, these two stipulations are extremely closely connected with a third, in which man figures as a subject operating egoistically.

All three of these seemingly incompatible stipulations which are not reducible to one and the same thing, express the totality of conditions necessary for the realisation of the value relation,

[1] Mikhail Ivanovich Tugan-Baranovsky, *Principles of Political Economy (Osnovy politicheskoy ekonomii)*, 1917, p. 60.

which is a relation in which social relations in the labour process appear as a reified characteristic of the products being exchanged.

The net result of abstracting these definitions from the actual social relation they express, and attempting to develop them as categories in their own right (by purely speculative means), is a confused jumble of contradictions and mutually exclusive propositions.[2] Yet in the exchange relation itself, these contradictions unite to form a dialectical totality.

The person engaged in exchange must be an egoist, that is to say he must stick to naked economic calculation, otherwise the value relation cannot be manifested as a socially necessary relation. The person engaging in exchange must be the bearer of rights, that is, he must be able to make autonomous decisions, for *his will* supposedly 'resides in objects'. Lastly, he embodies the principle of the essential equivalence of human personalities for, in exchange, all forms of labour are equalised and become human labour in the abstract.

Thus the three aspects mentioned above or, as people used to call them, the three principles of the egoism, freedom, and supremely equivalent worth of the personality are indivisibly linked and represent, in their totality, the rational expression of a single social relation. The egoistic subject, the legal subject and the moral personality are the three most important character masks assumed by people in commodity-producing society. The economics of value relations provides the key to an understanding of the juridical and ethical structure, not in the sense of the concrete content of legal or moral norms, but in the sense of the form itself. The idea of the worth and in

---

[2] The Jacobins, those petty-bourgeois revolutionaries, became tragically entangled in these mutually cancelling contradictions. They tried to subject the actual development of bourgeois society to the forms of civic virtue, forms borrowed from ancient Rome. On this, Marx has the following to say: 'What a terrible illusion it is to have to recognise and sanction in the *rights of man* modern bourgeois society, the society of industry, of universal competition, of private interest freely pursuing its aims, of anarchy, of self-estranged natural and spiritual individuality, and at the same time to want afterwards to annul the *manifestations of the life* of this society in particular individuals and simultaneously to want to model the *political head* of that society in the manner of *antiquity!*' (Karl Marx, 'The Holy Family', in: Marx and Engels, *Collected Works*, vol. IV, 1975, p. 122).

principle equal worth of the personality has a long history. It made the transition from Stoic philosophy to being employed by Roman jurists, went from there to the dogma of the Christian church, and thence to the doctrine of natural law. The existence of slavery in ancient Rome did not shake Seneca's conviction that 'even if the body be not free and belong to a master, yet the spirit always remains *sui juris*'. Basically, Kant made a very insignificant step forward as compared with this formula; in his work too, the fundamental autonomy of the personality can very conveniently be reconciled with purely feudal views of the relationship between master and servants. But regardless of the various forms this idea may have assumed, it expresses nothing but the fact that, as soon as the products of labour are exchanged as commodities, the different concrete types of socially useful labour are reduced to labour in the abstract. In all other relations, people's dissimilarity (sexual or class-determined) is so conspicuously apparent in the course of history that one is amazed, not by the profusion of arguments against the doctrine of people's natural equality put forward by its various opponents, but by the fact that, before Marx, no one had looked into the historical causes which produced this bias of natural law. For if, over the centuries, human thinking has returned with such persistence to the proposition that people are equal, and has elaborated this proposition in a thousand variations, then there must have been some objective reality behind it. The concept of the moral or equivalent personality is undoubtedly an ideological creation and, as such, not adequate to reality. Another equally ideological distortion of reality is the subject operating egoistically. Nonetheless, both these definitions are adequate to a specific social relation, it is just that they express this relation in an abstract and thus one-dimensional way. We have already had the opportunity of pointing out (in general terms) that the concept, or the little word 'ideology' should not deter us from further analysis. One would be making the task too easy for oneself if one were to set one's mind at rest with the idea that the person equal to all others is merely a product of ideology. 'Above' and 'below' are concepts which express only our 'earthly' ideology. Despite this, they are based on the undoubtedly real fact of gravity. Just as man recognised that the real cause of his com-

pulsion to differentiate between 'above' and 'below' was the
force of gravity directed towards the centre of the earth, he
also realised the limitation of these definitions, their inapplic-
ability to the whole cosmic reality. Thus the discovery of the
ideological nature of a concept was merely the reverse side of
establishing its accuracy.

If the moral personality is nothing but the subject in commodity-producing society, then the moral law must be manifested in the regulation of intercourse between commodity
owners. This inevitably endows the moral law with a dualistic
character. On the one hand this law must be a social law and
must therefore stand above the individual personality; on the
other hand, the owner of commodities is by nature the bearer
of a freedom (of the freedom, that is, to appropriate and to
alienate), which is why the rule governing transactions between
commodity owners must penetrate the soul of every commodity
owner, must be his inner law. Kant's categorical imperative
unites these contradictory requirements. It is supra-individual,
because it has nothing to do with natural inclinations at all,
with fear, sympathy, pity, the feeling of solidarity. According
to Kant, it neither intimidates, nor convinces, nor flatters. It
is beyond all empirical or purely human motives. At the same
time, it appears independently of any external pressure in the
direct, crude sense of the word. It is effective by virtue of the
consciousness of its universality. The Kantian ethic typifies
commodity-producing society, yet at the same time it is the
purest and most consummate form of ethic there is. Kant gave
a logically perfected shape to the form which atomised bourgeois society sought to embody in reality, by liberating the
personality from the organic fetters of the patriarchal and
feudal epochs.[3]

Hence the fundamental concepts of ethics lose their signific-

[3] It is very easy to combine Kant's ethical doctrine with belief in God,
particularly since it is the last refuge of this belief. For the two to be
linked is not, however, logically necessary. Furthermore, God, seeking
cover in the shadow of the categorical imperative, himself becomes a
flimsy abstraction scarcely suited to intimidating the masses. That is
why the feudal-clerical reaction sees it as its duty to polemicise against
Kant's lifeless formalism, to install their own, more reliable, as it were
'reigning' God, and to replace the categorical imperative with the living
feelings of 'shame, pity, and reverence' (V. Solovyov).

cance if considered in isolation from commodity-producing society, and if one attempts to apply them to any other social structure. The categorical imperative is by no means a social instinct, for its most important determinant is that it is effective where there is no possibility of organic, supra-individual motivation of any kind. Where there is a close emotional tie blurring the limits of the individual self, the phenomenon of moral obligation cannot occur. If one wants to comprehend this category, one must start out, not from the organic bond which exists, for example, between the mother animal and its young, or between the clan and each of its members, but from the condition of *isolation*. Moral being is a necessary complement of legal being; they are both modes of intercourse utilised by commodity-producers. The whole solemnity of Kant's categorical imperative comes down to the fact that man does 'freely', that is out of inner conviction, that which he would be compelled to do in the sphere of law. The examples Kant gives in illustration of his ideas typify this. Without exception, they are reduced to expressions of bourgeois propriety. There is no place for heroism and heroic deeds within the framework of the Kantian imperative. One is by no means obliged to sacrifice oneself, so long as one does not expect any such sacrifice of others. 'Irrational' acts of self-sacrifice and disregard for one's own interests for the sake of fulfilling one's individual historical destiny, one's individual social function, acts which stretch the social instinct to its limits, are beyond morality in the strict sense of the word.[4]

Schopenhauer, and subsequently Vladimir Solovyov, defined law as a certain ethical minimum. One would be equally justified in defining ethics as a certain social minimum. There is greater intensity of social feeling outside ethics in the strict sense of the word, as an inheritance left to the humanity of today by the organic system of past epochs, specifically by the gens-system. In comparing the nature of the ancient Germanic tribes with the civilised Romans, for instance, Engels says:

[4] Hence Professor Magaziner, for instance, is right to treat ethics in this sense as 'moderation and precision', and to contrast it with heroics which drive people to actions beyond their duty. (Cf. J. M. Magaziner, *General Theory of the State, Obshcheye uchenie o gosudarstve,* 2nd ed., Petrograd, 1922, p. 50).

Their personal efficiency and bravery, their love of liberty, and their democratic instinct, which regarded all public affairs as its own affairs, in short, all those qualities which the Romans had lost and which were alone capable of forming new states and of raising new nationalities out of the muck of the Roman world – what were they but the characteristic features of barbarians in the upper stage, fruits of their gentile constitution?[5]

The only respect in which rationalist ethics actually attains superiority over powerful, irrational, social instincts is its universality. It strives to burst all organic, necessarily narrow, bounds of the clan, the tribe, the nation and to become universal. It thereby reflects certain of mankind's material achievements, in particular the transformation of trade into world trade. The saying: 'neither Jew nor Greek' is an accurate reflection of a perfectly real situation in the history of the peoples united under the power of Rome.

The universalism of the ethical form (and thus also of the legal form) – the idea that all people are equal, possessing the same 'soul', that they all have the capacity to be legal subjects, and so forth – was forced on the Romans by the practice of trade with foreigners, that is with people of alien customs, alien language, alien religion. That is why it would scarcely have been regarded in a positive light in the first instance; if for no other reason than because it implied a renunciation of specific deep-rooted customs of their own such as love for their own and disdain for the alien. Thus Maine, for instance, points out that the *jus gentium* was itself a result of the low esteem in which the Romans held all foreign law, and of the fact that the Romans did not want the foreigners to partake of the advantages granted by their indigenous *jus civile*. According to Maine, the *jus gentium* was just as distasteful to the ancient Romans as it was to the foreigners for whom it was intended. The word *'aequitas'* itself meant levelling, and originally this expression probably contained no ethical nuance whatsover. There is no reason to suppose that the process indicated by this expression would have aroused any feeling other than repugnance in the

5 Friedrich Engels, 'The Origin of the Family, Private Property and the State', in: Marx and Engels, *Selected Works*, vol. III, 1970, p. 315.

primitive Roman mind.[6]

Nonetheless, the rationalist ethic of commodity-producing society subsequently presented itself as a great achievement and great cultural asset which it was usual to speak of only in tones of awe. One need only call to mind Kant's famous words:

> Two things fill the mind with ever new and increasing admiration and awe, the oftener and the more steadily we reflect on them: *the starry heavens above and the moral law within.*[7]

Meanwhile, when it is a matter of furnishing examples of such 'free' fulfilment of moral duty, the same old alms given a beggar, or refraining from a lie in circumstances where one could lie with impunity, and suchlike, are invariably trotted out. On the other hand, Kautsky remarks quite accurately that the maxim: 'treat your fellow-man as an end in himself', is meaningful in the situation where man can in practice be made another's means. Moral fervour is inextricably bound up with and feeds on the immorality of social practice. Ethical doctrines claim to change and improve the world whereas, in reality, they are merely a distorted reflection of one aspect of this real world – the aspect which shows social relations to be subject to the law of value. It should not be forgotten that the moral personality is only one of the three *hypostases* within one subject; man as an end in himself is only another aspect of the subject operating egoistically. An act which is the real, and only true, embodiment of the ethical principle simultaneously also contains its negation. The large capitalist destroys the small capitalist 'in good faith', without thereby violating the absolute worth of his personality in any way. The proletarian's personality is 'equal in principle' to that of the capitalist; this circumstance is expressed in the fact of the 'free' contract of employment. But all that comes out of this 'materialised freedom' for proletarians is the freedom to die of starvation – without any intereference.

This ambiguity of the ethical form is not fortuitous; nor is it an external defect conditioned by the specific shortcomings of

[6] Henry Summer Maine, *Ancient Law* (1861), 10th ed., London: John Murray, 1905, pp. 50 and 60. [Ed.]
[7] *Kant's Critique of Practical Reason*, translated by T. K. Abbott, 3rd ed., London: Longmans, Green & Co., 1883, Conclusion, p. 260.

capitalism. On the contrary, it is a characteristic feature of the ethical form as such. Eliminating the ambiguity of the ethical form implies the transition to planned, socialised economy. This in turn means the construction of a social system which enables people to build and conceive of their interrelations in terms of the clear and simple concepts of harm and advantage. Abolition of the ambiguity of the ethical form in the most substantive area, that is in the sphere of material existence, implies abolishing the ethical form altogether.

In its attempt to disperse the metaphysical mists surrounding ethical doctrine, pure utilitarianism considers the concepts of 'good' and 'evil' from the point of view of harm and advantage. In so doing, it in fact abolishes ethics or, more accurately, it attempts to invalidate and to abolish it. For the abolition of moral fetishes can only be accomplished in practice simultaneously with the abolition of commodity fetishism and legal fetishism.[8] Until humanity has reached this level of historical development, that is, until the heritage of the capitalist epoch has been transcended, the efforts of theoretical elaboration will only anticipate this coming liberation, but cannot embody it in practice. We would like to call to mind Marx's words on commodity fetishism:

> The belated scientific discovery that the products of labour, in so far as they are values, are merely the material expressions of the human labour expended to produce them, marks an epoch in the history of mankind's development, but by no means banishes the semblance of objectivity possessed by the social characteristics of labour.[9]

But, people will retort, the class morality of the proletariat is already in the process of liberating itself from all fetishes. Moral obligation is that which is useful to the class. There is nothing absolute about morality in a form such as this, for what is advantageous today may cease to be of use tomorrow; nor is there anything mystical or supernatural about it, since

---

[8] The German edition omits the following sentence from the 3rd Russian edition: 'People whose conduct is guided by the clear and simple concepts of harm and advantage will not require that their social relations be expressed in terms of value or in terms of law'. [Ed.]
[9] Marx, *Capital*, vol. I, 1976 ed., p. 167.

the utilitarian principle is simple and rational.

There is no doubt that the morality of the proletariat, or rather the morality of its avant-garde, loses its doubly fetishistic character by being purged of religious elements. Even a morality free of any impurity in the form of religious elements is nevertheless still morality, that is to say it is a form of social relations in which everything has not yet been reduced to man himself. If the living bond linking the individual to the class is really so strong that the limits of the ego are, as it were, effaced, and the advantage of the class actually becomes identical with personal advantage, then there will no longer be any point in speaking of the fulfilment of a moral duty, for there will then be no such phenomenon as morality. However, where such a fusion has not occurred, the abstract relation of moral duty with all the forms arising from that, will inevitably occur. The precept: 'act in such a way that you are of the greatest possible use to your class' will then sound exactly the same as Kant's formula: 'act only on that maxim through which you can at the same time will that it should become a universal law'.[10]

The only difference is that in the first case we make a concrete reservation in giving ethical logic a class-orientated framework.[11] Within this framework, however, it retains its full significance. The class content of morality does not of itself destroy its form. We have in mind here not only the logical form, but also the actual form in which it occurs. Even within a proletarian, or class-orientated, collective we may observe the same formal methods of bringing the moral imperative to bear, methods which are composed of two contradictory motives. On the one hand the collective does not relinquish every poss-

---

[10] *The Moral Law or Kant's Groundwork of the Metaphysics of Morals*, edited and translated by H. J. Paton, London: Hutchinson's University Library (1948), 3rd ed., 1956, p. 88; (p. 52 in the 2nd German edition). [Ed.]

[11] It is self evident that, within a society torn by class struggles, a classless ethic can exist only in the imagination, but not in practice. The worker who decides to take part in a strike, regardless of the privations he will suffer as a consequence of this decision, is entitled to see his decision as the moral duty to subordinate his own private interests to the common interest. Yet it is quite obvious that this conception of the common interest cannot encompass the interests of the capitalist against whom the labour struggle is being waged.

ible means of pressurising its members into fulfilling their moral duty. On the other hand, the same collective only characterises conduct as moral when there is no such external pressure motivating it. This explains precisely why, in social practice, morality and moral conduct are so closely linked with hypocrisy. It is true that the proletariat's conditions of existence contain the prerequisites for the development of a new, higher, more harmonious form of link between the personality and the collective. There are many examples of the expression of proletarian class solidarity which illustrate this. Yet the old continues to exist alongside the new. Beside the social person of the future, who submerges his ego in the collective and finds the greatest satisfaction and the meaning of life in this act, the moral person too still persists, bearing on his shoulders the burden of a more or less abstract duty. The victory of the first form is synonymous with total liberation from all vestiges of private-law relations, and with the ultimate transformation of humanity in the light of the ideas of communism. Of course this task is by no means purely ideological or paedagogical. The new type of social relations requires the creation and consolidation of a new material, economic base.

One must, therefore, bear in mind that morality, law and the state are forms of bourgeois society.

The proletariat may well have to utilise these forms, but that in no way implies that they could be developed further or be permeated by a socialist content. These forms are incapable of absorbing this content and must wither away in an inverse ratio with the extent to which this content becomes reality. Nevertheless, in the present transition period the proletariat will of necessity exploit this form inherited from bourgeois society in its own interest. To do this, however, the proletariat must above all have an absolutely clear idea -- freed of all ideological haziness -- of the historical origin of these forms. The proletariat must take a soberly critical attitude, not only towards the bourgeois state and bourgeois morality, but also towards their own state and their own morality. Phrased differently, they must be aware that both the existence and the disappearance of these forms are historically necessary.[12]

12 Does this mean, then, that 'there will be no morality in the society of the future'? Not at all, if one understands morality in the wider sense

In his critique of Proudhon, Marx points out that the abstract concept of justice is by no means an absolute and eternal criterion on the basis of which one can erect an ideal, or rather, a just, exchange relation. This would represent an attempt

> to metamorphose chemical combustion in line with 'eternal ideas', 'particular qualities' and 'affinities', instead of studying the laws actually governing it.[13]

For the concept of justice is itself inferred from the exchange relation and has no significance beyond this. Basically, the concept of justice does not contain anything substantively new, apart from the concept of the equal worth of all men which we have already analysed. Consequently it is ludicrous to see some autonomous and absolute criterion in the idea of justice.

That is not to deny that this idea, skilfully applied, lends itself admirably to interpreting inequality as equality, and hence to veiling the ambiguity of the ethical form. At the same time, justice is the stage through which ethics passes to become law. Moral conduct must be 'free', but justice can be enforced. Compulsion in moral behaviour attempts to deny its own existence; in contrast to this, justice is 'allotted' to man openly; it admits of superficial execution and active egoistic interest. These are the most important points of contact and divergence of the ethical form and the legal form.

Exchange, or the circulation of commodities, is predicated on the mutual recognition of one another as owners by those engaged in exchange. This acknowledgement, appearing in the form of an inner conviction or of the categorical imperative, is

---

as the development of higher forms of humanity, as the transformation of man into a species-being (to use Marx's expression). In the given case, however, we are talking about something different, about specific forms of moral consciousness and moral conduct which, once they have played out their historical role, will have to make way for different, higher forms of the relationship between the individual and the collective. (Note to the 3rd Russian edition.) 'Species-being' is a term taken by Marx from L. Feuerbach and used by him in his early writings (e.g. in the passage quoted above, at the end of chapter 4), especially the *"Economic and Philosophical Manuscripts 1844"* (in Marx and Engels, *Collected Works* Vol. III)—Ed.]

[13] Marx, 'The Poverty of Philosophy', in: Marx and Engels, *Collected Works*, vol. VI, 1976, is presumably what Pashukanis refers to here, but we have been unable to identify the exact citation. [Ed.]

the maximum conceivable height to which commodity-producing society can rise. But in addition to this maximum there is also a certain minimum condition under which commodity circulation can still operate freely. For this minimum to be realised, it is sufficient for commodity owners to act *as if* they acknowledged one another mutually as proprietors. Moral conduct is here contrasted with legal conduct, which is characterised as such irrespective of the motives which generated it. From the juridical standpoint, it makes no difference whether the debt is repaid because 'one would in any case be forced to pay up', or because the debtor feels morally obliged to pay. The idea of external coercion and, beyond this, the organisation of that coercion, are important elements of the legal form. If legal intercourse can be construed, in purely theoretical terms, as the reverse of the exchange relation, then its realisation presupposes the existence of more or less fixed general patterns, an elaborate casuistry and, finally, a particular organisation which applies these patterns to individual cases and sees to the compulsory execution of sentences. These needs are best fulfilled by the state power, although legal intercourse often manages even without its support, using the law of custom, voluntary arbitration, self-help, and so on.

In situations where the function of coercion is not organised, and not assigned to a special apparatus standing above the parties, it figures in the shape of so-called 'reciprocity'. Under the condition of equilibrium of forces, this principle of reciprocity represents the only, and, it must be said, extremely insecure basis of international law to this very day.

On the other hand, the legal claim appears – contrary to the moral claim – not in the shape of an 'inner voice', but as an external claim emanating from a concrete subject who is also, as a rule, simultaneously the bearer of a corresponding material interest.[14] That is why the fulfilment of legal obligation is alien

---

[14] This is how it is in private law, which is the general prototype of the legal form. 'Legal claims' emanating from organs of public power, claims without any private interest behind them, are nothing more than juridical stylisations of the facts of political life. The character of these stylisations differs according to the circumstances, which is why the juridical conception of the state is undeniably prone to pluralism. If state power is posited as the embodiment of an objective rule standing above the parties-cum-subjects, it merges, as it were, with the norm and becomes

to all subjective elements in the person being obliged in this way and takes the external, almost material, form of the *fulfilment of a claim*. As a result, the concept of legal obligation itself becomes very problematic. If one is absolutely consistent, one must say – as Binder does in his *Legal Norm and Legal Obligation* – that a legal obligation (*rechtliche Verpflichtung*) has nothing in common with 'duty' (*Pflicht*), but exists juridically only as 'liability' (*Haftung*); 'to be liable' (*verpflichtet*) means nothing but

> to answer with one's property (and in criminal law with one's person as well); the liability persists in, and consists of, the trial and the enforcement of its outcome.[15]

Most jurists find Binder's conclusions – which can be expressed in the brief formula of 'law does not legally commit one to anything' ('*das Recht verpflichtet rechtlich zu nichts*)' – paradoxical, yet in reality they are merely the consistent extension of that division of concepts begun by Kant. But precisely this strict differentiation of the moral and the juridical sphere is the source of insoluble contradiction for the bourgeois philosophy of law. If legal obligation has nothing in common with 'inner' moral duty, then there is no way of differentiating between subjection to law and subjection to authority as such. Yet if one admits that the aspect of duty is an important feature of law, even if with only the faintest subjective tint imaginable, then the significance of law as a socially necessary minimum is immediately lost. Bourgeois legal philosophy

---

highly impersonal and abstract. The state's claim appears as an impartial, disinterested statute. In this situation it is almost impossible to conceive of the state as a subject, so entirely does it lack substance, so thoroughly has it become the abstract guarantee of intercourse between real subjects-cum-commodity owners. This view, as the most unadulterated juridical conception of the state, is the one championed by the Austrian normative school, led by Kelsen.

In international dealings, to the contrary, the state in no way appears as the personification of an objective norm, but figures rather as the bearer of subjective rights, that is to say with every attribute of substantiality and of egoistic self-interest. The state plays this same role when it functions, in its capacity as treasury, as a party in litigation with private individuals. There are numerous plausible intermediate and hybrid forms between these two conceptions of the state.

[15] J. Binder, *Rechtsnorm und Rechtspflicht*, Leipzig, 1912.

exhausts itself in this fundamental contradiction, this endless struggle with its own premises.

The interesting point about this is that the basically identical contradictions are manifested in two different forms, according to whether it is the relationship between law and morality, or between law and the state, that is the point at issue. In the first case, if the independence of law from morality is being asserted, law merges with the state as a result of the strong emphasis on the aspect of external coercion. In the second case, if law is being contrasted with the state, that is to say with effective dominance, then the aspect of duty in the sense of Ought (Sollen, as distinct from Müssen) comes into play without fail, and we are confronted – if one may put it like this – by a united front of law and morality.

Petrazhitsky's attempts to come up with an imperative for law which was absolute, that is, ethical, yet at the same time differed from the moral Ought, have remained fruitless.[16] As we well know, Petrazhitsky construes the category of legal obligation as a duty incumbent on someone and owed to someone else, subject to claims on us by that person. In contrast to this, moral obligation in his opinion merely prescribes a certain conduct without suggesting that third persons should demand their due. It follows that law has a bilateral, imperative-attributive character, whilst morality is unilaterally binding, or purely imperative. Petrazhitsky's argument is based on introspection. He assures us that he can distinguish effortlessly between legal obligation, which leads him to refund his creditor the amount he borrowed, and the moral duty which occasions his giving alms to a beggar. It appears, however, that this ability to distinguish so clearly is peculiar to Professor Petrazhitsky. For others, such as Trubetskoy,[17] assure us that the obligation to give a beggar alms is just as directly related to the beggar, in psychological terms, as is the obligation to pay back the debt to the creditor. (A thesis which – moreover – is not disadvan-

[16] Lev Yosifovich Petrazhitsky, Introduction to the Study of Law and Morality (Vvedenie v izuchenie prava i nravstvennosti); [abridged form translated by H. W. Babb, as: Law and Morality, with an introduction by N. S. Timasheff, Cambridge, Mass: Harvard University Press, 1955 (20th Century Legal Philosophy Series, vol. 7). Transl.]
[17] E. Trubetskoy, Encyclopaedia of Law (Entsiklopediya prava), Moscow, 1908, p. 28.

tageous to the beggar, but must surely appear very questionable to the creditor). In contrast to this, Reisner is of the
opinion that the emotion of a prescribed obligation appertains
entirely to the psychology of power. Thus, if Trubetskoy puts
the creditor with his claim on the same footing as the beggar,
'psychologically' speaking, then Reisner places the creditor in
a position of authority. To put it another way: the contradiction which we have shown, in a logical and systematic manner,
to be a contradiction in concepts, occurs here as a contradiction
in the results of introspection. Yet its significance remains unaltered. Legal obligation can find no independent validity and
wavers interminably between two extremes: subjection to
external coercion, and 'free' moral duty.

As always, the contradiction in the system here too reflects
the contradictions in real life, that is in the social environment
which produced the form of morality and law as they exist. The
contradiction between the individual and the social, between
the private and the universal, which bourgeois philosophy is
unable to do away with, despite all its efforts, is the very basis
of life in bourgeois society as a society of commodity producers. This contradiction is embodied in the actual interrelations of people who cannot regard their private endeavours as
social aspirations except in the absurb and mystified form of the
value of commodities.

7. Law and the Violation of Law

The *Russkaya Pravda*, that oldest juridical memorial of the Kiev period of our history, has all in all amongst its forty-three articles[1] (the so-called 'academic list') only two articles which do not refer to violations of criminal or civil law. All remaining articles either define sanctions, or contain procedural rules to be applied in case of offences against the law. Hence deviation from the norm is a prerequisite in both cases.[2] The so-called barbaric laws of the Germanic tribes reveal a similar picture. Thus, for instance, of the 408 articles of Salic law, only 65 do not have a penal character. The most ancient memorial of Roman law, the laws of the twelve tablets, begin with the regulation dealing with the summons before court: '*Si in jus vocat, ni it, antestamino igitur in capito*'.[3] The well-known historian Maine says in his book, *Ancient Law*:

> It may be laid down, I think, that the more archaic the code, the fuller and the minuter is its penal legislation.[4]

[1] Cf. *Medieval Russian Laws*, translated by G. Vernadsky (Columbia University Press, 1947), reprinted New York: Octagon Books Inc., 1965, pp. 26-35: *The Short Version*, including 'Yaroslav's *Pravda*' (Articles 1-18) and the '*Pravda* of Yaroslav's Sons' (Articles 19-43). [Ed.]

[2] There is surely no need to make specific mention of the fact that, at this primitive stage of development, no distinction was made between criminal and civil 'wrong'. The concept of injury requiring compensation was dominant: theft, robbery, murder, failure to repay a debt, were without exception regarded as grounds on which the injured party could bring an action and receive satisfaction in the form of a money fine.

[3] XII tablic., edited by Nikol'sky, 1897, p. 1. [Ed.]

[4] Henry Summer Maine, *Ancient Law* (1861), 10th ed., London: John Murray, 1905, p. 368.

Non-acquiescence to the norm, violation of the norm, rupture of normal intercourse and resulting conflict, is the point of departure and main content of archaic legislation. Yet the normal as such is not prescribed at first; it simply does not exist. The need to fix and determine precisely the extent and the content of mutual rights and obligations first arises when calm, peaceful existence is disrupted. From this point of view, Bentham is right in saying that law creates right by creating crime. Historically, the specific traits of legal intercourse were acquired primarily as a result of actual violation of the law. The concept of theft arose before the concept of property. The relations arising out of loan are fixed to cover possible default on the part of the debtor: 'if a man claims repayment of a debt from someone else, and the latter baulks', and so forth.[5]

The original meaning of the word '*pactum*' is not at all contract in general, but is derived from '*pax*' (peace), that is to say it means the peaceful termination of a dispute: the accord marks an end of discord (*der Vertrag bereitet der Unverträglichkeit ein Ende*).[6]

Accordingly, if private law reflects, in the most direct form, the general conditions of existence of the legal form as such, then criminal law is the sphere in which legal intercourse is most severely tested. It is here that the juridical element first and most crudely detaches itself from everyday life and becomes fully autonomous. The transformation of the actions of a concrete person into the proceedings of a legal party, that is of a legal subject, is particularly apparent in the court case. In order to distinguish everyday acts and expressions of will from juridical expressions of will, ancient law utilised special, solemn formulae and ceremonies. The drama of court proceedings graphically created a peculiarly juridical reality, parallel with the real world.

Of all types of law, it is precisely criminal law which has the capacity to affect the individual person in the most direct and unmitigated manner. That also explains why the most intense practical interest has always been focussed on criminal law.

---

[5] *Russkaya pravda*, 'Academic List', Article 15, in: *Medieval Russian Laws*, op. cit., p. 29. [Ed.]
[6] Cf. Rudolf von Ihering, *Geist des römischen Rechts auf den verschiedenen Stufen seiner Entwicklung*, Russian translation, 1875, vol. I.

The law and punishment for its infringement are extremely closely linked, so that criminal law plays, as it were, the role of a representative of law as such; it is a part which stands for the whole.

Historically, the origin of criminal law is associated with the custom of blood vengeance. Undoubtedly these two phenomena are genetically very close to one another. However, vengeance *really becomes vengeance* only because it is followed by fines and sentences; here too it is only the subsequent stages of development (as is so often to be observed in the history of mankind) which render comprehensible the implications contained in the preceding forms. If one approaches the same phenomenon from the opposite end, one can see nothing in it but the struggle for existence, a purely biological fact. For the criminologists, fixated on a later epoch, blood vengeance is synonymous with the *jus talionis*, that is, with the principle of equivalent requital, whereby the possibility of further vengeance is excluded if the injured party or his clan have avenged the offence. In reality, as Kovalevsky rightly demonstrates, the earliest nature of blood vengeance was completely and utterly different. Clan strife is passed on from generation to generation. Every offence, even that perpetrated in revenge, forms grounds for a new blood vengeance. The injured party and his clansmen become, in turn, those giving offence, and so it goes on from generation to generation, often to the point of total annihilation of the warring clans.[7]

Vengeance first begins to be regulated by custom and becomes transformed into retribution according to the rule of the *jus talionis*: 'an eye for an eye, a tooth for a tooth', at the time when, apart from revenge, the system of compositions or of expiatory payment is adopted. The idea of the equivalent, this first truly juridical idea, itself originates in the commodity form. Felony can be seen as a particular variant of circulation, in which the exchange relation, that is the contractual relation, is determined retrospectively, after arbitrary action by one of the parties. The ratio between offence and retribution is like-

[7] Cf. M. Kovalevsky, *Modern Custom and Ancient Law (Sovremenny obychay i drevny zakon)*, Petersburg and Moscow, 1886, vol. II, pp. 37 and 38; [translated as: *Modern Custom and Ancient Law of Russia*, London: D. Nutt, 1891. Trans.]

wise reduced to this exchange ratio. That is why Aristotle, when speaking of equalisation in exchange as a form of justice, divides this into two sub-species, equalisation by voluntary and by involuntary acts. In the process, he considers the economic relations of buying, selling, loans, and so forth as the former, and all forms of crime which entail a punishment as their equivalent as the latter. He is also the author of the definition of crime as an involuntarily concluded contract. Punishment emerges as an equivalent which compensates the damage sustained by the injured party. As we know, this idea was also adopted by Hugo Grotius. Naïve as these constructs may seem at first sight, they conceal a much finer feeling for the legal form than is to be found in the eclectic theories of modern jurists. In the examples of vengeance and punishment, we can observe with particular clarity the imperceptible nuances by which the organic and biological is related to the juridical. This fusion is heightened by the fact that man is not able to renounce the interpretation of the phenomena of animal life to which he is accustomed, that is the juridical (or ethical) interpretation. In the actions of animals, man unintentionally finds the significance which was actually only projected onto them by subsequent development, that is, by the historical development of humanity.

Self-defence is one of the most natural phenomena of animal life. It is completely irrelevant whether we find it only as an individual reaction of a lone creature, or as a reaction of a collective. According to the evidence of scholars who study the life of bees, the bees guarding the entrance to the beehive attack and sting any bee not belonging to the swarm if it attempts to gain entry to the hive in order to steal honey. Moreover, if a bee from another swarm has already gained access to the hive, it is killed the instant it is discovered. It is not unusual, in the animal world, to find cases where the reaction is separated by a certain interval of time from the action causing it. The animal does not respond immediately to the attack, but delays this until a later, more suitable time. Self-defence here becomes vengeance in the truest sense of the word. And since, for modern man, vengeance is inextricably linked with the idea of equivalent retribution, it therefore comes as no great surprise that Ferri, for example, wishes to assume the existence of a 'juridical' instinct in animals.[8]

In fact, the juridical idea, or the idea of the equivalent, is first clearly delineated and objectively realised at that level of economic development where this form becomes common as equalisation in exchange. This certainly does not occur in the animal world, but in human society. It is entirely unnecessary to this process for expiatory payment to have completely supplanted vengeance. It is precisely in cases where restitution has been refused as something dishonourable (a view prevalent for a long while among primitive peoples), and where the carrying out of personal revenge was regarded as a solemn duty, that the act of vengeance acquires a new nuance which it did not have when it was not yet an alternative. Now, the idea of vengeance as the only adequate retribution is projected onto it. The renunciation of restitution in money form emphasises, so to speak, that the shedding of blood is the only equivalent for blood already shed. Vengeance is transformed from a purely biological phenomenon into a juridical institution by being linked with the form of equivalent exchange, exchange according to values.

Archaic penal law emphasises this connection in a particularly crude and vivid manner, in that it directly equates the damage done to property and injury to the person with a *naïveté* which later epochs shamefacedly renounced. From the standpoint of ancient Roman law, there was nothing strange in the fact that a tardy debtor paid for his debts with parts of his body (*in partes secare*), and that the person guilty of bodily harm paid for it with his property. The idea of the equivalent transaction appears here in all its bluntness, uncomplicated and unsoftened by any additional aspects. In accordance with this, criminal procedure assumes the character of a commercial transaction.

We have here to visualise a transaction in which one side made suggestions and the other bargained, until finally they came to terms. The expression for this was *pacere, pacisci, depecisci*, and for the agreement itself *pactum*. Here begins the role of the mediator in ancient Nordic law, chosen by both parties, who determines the sum of money paid as compensation (the *arbiter* in the original Latin sense).[9]

---

[8] Enrico Ferri, *Sociologia criminale*, Russian translation with a preface by Drill', vol. II, p. 37.
[9] Ihering, *Geist des römischen Rechts*, ed. cit., vol. I, p. 136.

As far as so-called public penalties are concerned, there is no doubt that they were originally introduced primarily for fiscal considerations and served as a means of filling the coffers of the representatives of power. Maine has this to say about it:

> The State did not take from the defendant a composition for any wrong supposed to be done to itself, but claimed a share in the compensation awarded to the plaintiff, simply as the fair price of its time and trouble.[10]

We know from Russian history that this 'fair compensation for time lost' was exacted so assiduously by the princes that, according to the testimony of the chronicler, 'the Russian earth was ravaged by wars and penal payments'. The same phenomenon of judicial plundering can be observed, not only in the old Russia, but also in the Empire of Charlemagne. In the eyes of the princes of ancient Russia, judicial fees were in no way different from their other sources of revenue. They made presents of them to their servants, divided them up, and so on. It was possible to buy oneself free from the princely court by paying a certain sum.[11]

Besides public penalty as a source of income, punishment as a means of maintaining discipline and protecting the authority of sacerdotal and military power emerged fairly early. It is well-known that, in ancient Rome, most serious offences were simultaneously crimes against the gods.[12] Thus a violation of law which was very important to landowners, such as the wilful displacement of boundary markers, was traditionally viewed as a religious offence, and the guilty person's head fell into the power of the gods. The priestly caste, who acted as the custodians of order, thereby pursued not merely an ideal, but also a very solid material interest, for the fortune of the guilty party

---

[10] Maine, *Ancient Law*, ed. cit., p. 378.

[11] Cf. the 'vira' (bloodwite) of the *Russkaya pravda*, in: *Medieval Russian Law*, op. cit., notes to Article 19 of the *Short Version*, p. 30; and to Articles 3-8 of the *Expanded Version*, pp. 36-37. [Transl.]

[12] Since the oath *juramentum* was the most indispensable part of legal intercourse (Ihering thinks that 'to bind oneself', 'to establish a right', and 'to swear', were for a long period considered as synonymous), all legal intercourse was placed under religious protection, for the oath itself was a religious act, and perjury was a religious offence. (Cf. Ihering, *Geist des römischen Rechts*, op. cit., p. 304).

was confiscated in their favour in such cases. In other circumstances, even penalties inflicted by the priestly caste for attacks on their revenue – refusal of prescribed ceremonies or of sacrificial offerings, attempts to introduce new religious doctrines of any sort, and so forth – were of a public nature.

The influence of priestly organisations, that is, of the church, on criminal law is illustrated by the fact that, although the sentence still retained the character of an equivalent or a *retribution*, this retribution is no longer directly linked with the loss to the injured party based on his claim, but acquires a higher, abstract significance as divine punishment. In this way, the church attempts to associate the ideological motive of atonement (*expiatio*) with the material aspect of compensation for the injury, and thus to construct, from penal law based on the principle of private revenge, a more effective means of maintaining public discipline, that is to say class rule. The efforts of the Byzantine clergy to have capital punishment introduced in the principality of Kiev are characteristic of this. The same end, the maintenance of discipline, also determined the nature of the penal measures of military leaders. They sat in judgement over subjected people as well as over their own soldiers for plotting military treason or simply for insubordination. The famous story of Clovis, who cracked open the head of a recalcitrant warrior with his own two hands, shows how primitive this court was in the days when the barbarian kingdoms of the Germanic tribes were being established. In earlier days, it was the assembly of the people which had the task of maintaining military discipline. With the strengthening and stabilisation of royal power, this function naturally transferred to the kings and became identical with the defence of their own privileges. As for the remaining criminal offences, for a long period the Germanic kings (as also the princes of Kiev) had a purely fiscal interest in them.[18]

---

[18] We know that in ancient Russian law, the expression 'self-help' meant primarily that the court costs due to the prince were withheld from him. Similarly, in King Erik's statute-book, private settlements between the injured party or his relatives and the offender are strictly forbidden, if the share due to the king should thereby be withheld from him. In the same Codes, however, permission to bring an action in the name of the king or the bailiff is very rare. (Cf. Wilhelm Eduard Wilda, *Das Strafrecht der Germanen*, vol. I of: *Geschichte des deutschen Strafrechts*, 1842, p. 219).

This state of affairs changes with the development and stabilisation of barriers of rank and class. The emergence of a spiritual and a temporal hierarchy highlights the protection of their privileges and the struggle against the lower, oppressed classes of the population. The dissolution of natural economy and the increased exploitation of the peasants which resulted, the evolution of trade and the organisation of the state based on rank and class confront criminal justice with entirely new problems. Criminal justice in this epoch is no longer simply a means for those in power to fill their coffers, but is a means of merciless and relentless suppression, especially of peasants fleeing intolerable exploitation by lords of the manor and the seigneurial state, of impoverished vagabonds, beggars, and so forth. The police and the investigative apparatus begin to assume the most important function. Penalties become the means of either physical extermination or intimidation. It is the epoch of torture, of capital punishment, of gruesome forms of execution.

Thus the way was gradually prepared for the complex amalgam of modern criminal law. It is a simple matter to distinguish between the historical strata from which it emerged. Basically, that is to say from the purely sociological standpoint, the bourgeoisie maintains its class rule and suppresses the exploited classes by means of its system of criminal law. In this respect, its courts and its private, 'voluntary' organisations of strike-breakers are pursuing one and the same end.

Considered in this light, criminal justice is merely an adjunct of the investigative and police apparatus. Should the Paris courts one day close their portals for a few months, the only people to suffer would be those offenders already arrested. If, on the other hand, the notorious police brigades of Paris were to stop work for just one day, the result would be catastrophic.

Criminal justice in the bourgeois state is organised class terror, which differs only in degree from the so-called emergency measures taken in civil war. Spencer long ago drew attention to the complete analogy between, indeed the identity of, defensive action against an external attack (war) and the reaction against disruptive elements in internal state organisation (judicial or juridical defence).[14] The fact that measures of the first kind,

[14] Herbert Spencer, *The Principles of Sociology* (1876), 2 vols., London and Edinburgh: Williams and Norgate, 1893; vol. I, pp. 565-575.

that is penal measures, are applied chiefly against elements who have lost their position in society, and that measures of the second kind are *mainly* employed against the most active fighters of a new class rising to power, alters the principle behind the matter just as little as the greater or lesser uniformity and complexity of the procedure employed. One can only grasp the true significance of the penal practice of the class state by starting from its antagonistic nature. The would-be theories of criminal law which derive the principles of penal policy from the interests of society as a whole are conscious or unconscious distortions of reality. 'Society as a whole' does not exist, except in the fantasy of the jurists. In reality, we are faced only with classes, with contradictory, conflicting interests. Every historically given system of penal policy bears the imprint of the class interests of that class which instigated it. The feudal lord had intractable peasants and townspeople who opposed his power executed. The confederate cities hung the robber knights and destroyed their strongholds. In the Middle Ages, every person who tried to follow a trade without being a member of the guild was thought to be a law-breaker. The capitalist bourgeoisie, scarcely had it emerged, declared that the workers' attempts to join forces in associations were criminal.

Thus class interest impresses the stamp of historical concreteness on every system of penal policy. As regards the individual methods of penal policy, people usually point to the great progress made by bourgeois society towards more humane forms of punishment since the days of Beccaria and Howard.[15] The abolition of torture, of corporal punishment and

[15]John Howard, 1726-90, the English prison reformer, toured all Europe urging prison reform and especially urging the payment of wages to gaolers who had lived previously on fees extracted from those in their charge. The publication of his book *The State of the Prisons* in 1777 coupled with his other efforts led to new legislation.

Angelicus Beccaria was the pseudonym for Richard Wright, a Unitarian missionary. In 1807 he wrote his *Letters on Capital Punishment to the English Judges*. They begin: 'I have undertaken to prove that capital punishments are unnecessary, useless and injurious: that we have no authority . . . to put people to death.'

But Beccaria's devotion to the protection of property rather abruptly calls into question the universality of the divine imperatives which on other occasions he invokes in his efforts to remove, or at least to limit the use of capital punishments: 'If the peace and good order of society could not be preserved and property not be protected, in a word, if all

humiliating punishments, of gruesome forms of execution, and so on, are relevant here. There is no doubt that all this represents a great step forward. However, one should not forget that the abolition of corporal punishment has not been by any means universally achieved. In England, flogging with a birch is permissible – up to twenty-five lashes for minors under sixteen years of age; up to 150 lashes for adults – as punishment for theft and plundering. The cat-o'-nine-tails is used on sailors in England. In France, corporal punishment is used as a disciplinary punishment for prison inmates.[16] In America, in two states of the union they mutilate offenders by castrating them. Denmark introduced flogging with a rod and with tarred rope-ends for a number of offences in 1905. Only a short time ago, the fall of the Soviet Republic in Hungary was celebrated by the introduction of, amongst other things, caning of adults for a whole series of offences against the person and against property.[17] Furthermore, it is worthy of note that precisely the last decades of the nineteenth century and the first decades of the twentieth century have seen a perceptible trend in a whole number of bourgeois states towards reintroduction of excruciatingly painful and humiliating punishments as deterrents. The humanism of the bourgeoisie is replaced by the demand for severity and the more frequent application of the death sentence.

Kautsky tries to explain this by the fact that, at the end of the eighteenth and the beginning of the nineteenth centuries, in other words until the introduction of universal military service, the bourgeoisie had a peaceful and humane disposition because it did not serve in the army. Surely this can scarcely be the main reason. More important is the transformation of the bourgeoisie into a reactionary class, fearing the rise of the working-class movement, and, lastly, the policy of colonialism, which has been a school of cruelty all along.

Only the complete disappearance of classes will make possible the creation of a system of penal policy which lacks any element of antagonism. Moreover, it is very doubtful whether,

---

the ends of human justice could not be secured without the infliction of capital punishments, not a single argument ought to be advanced against their continuance.' [Ed.]

[16] Cf. Ivan Yakovlevich Foinitsky, *The Theory of Punishment* (*Uchenie o nakazanii*), p. 15.

[17] Cf. *Deutsche Strafrechtszeitung*, 1920, nos. 11-12.

in such circumstances, there will be any necessity at all for a penal system. If the penal practice of the state power is *by nature* and *in its content* a weapon for the protection of class rule, then it will appear *in its form* as an aspect of the legal superstructure, and will be absorbed into the legal system as one of its branches. We have already shown that the naked struggle for existence takes on juridical form through the projection into it of the principle of equivalence. Thus the act of self-defence ceases to be purely self-defence and becomes a form of exchange, a peculiar form of circulation, which has its place alongside 'normal' commercial circulation. Crime and punishment become what they are, in other words they take on a juridical stamp, through a buying-off transaction. To the extent that this form is retained, the class struggle takes place in the form of the administration of justice. Conversely, the characterisation 'criminal law' becomes utterly meaningless if this principle of the equivalent relation disappears from it.

Hence, criminal law, in so far as it is a variation of that basic form to which modern society is subject – precisely the form of equivalent exchange with all its consequences – becomes a constituent part of the legal superstructure. The materialisation of this exchange relation in criminal law is one aspect of the constitutional state as the embodiment of the ideal form of transaction between independent and equal commodity producers meeting in the market. However, since social relations are not confined to the abstract relations between abstract commodity owners, so too the criminal court is not only an embodiment of the abstract legal form, but is also a weapon in the immediate class struggle. The sharper and more bitter this struggle, the more difficult it will be for class rule to be realised within the legal form. When that happens, the 'impartial' court and its legal guarantees will be ousted by an organisation of unmediated class violence, with methods guided by considerations of political expediency alone.

If one regards bourgeois society as a society of commodity owners, one would assume *a priori* that its criminal law would be 'most juridical' in the sense illustrated above. Yet we appear here to be confronted with various difficulties right from the start. The first difficulty is that modern criminal law starts out, not at all from the damage suffered by the injured party, but from the violation of the norm established by the state. Yet if the injured party with his claim steps into the background, one might well ask what then has become of the equivalent form. Nevertheless, in the first place, however much the injured party may fade into the background, he does not disappear altogether. He continues to signify the context of the criminal action taking place. The abstraction of injured public interest is based on the perfectly real figure of the injured party, who takes part in the trial either personally or through a representative and gives the trial its living meaning.[18] Furthermore, this abstraction is personified in a real way even in those cases where there is in fact no injured party, and only the law 'starts up in anger' in the person of the public prosecutor.

This split, whereby state power appears not only in the role of plaintiff (public prosecutor), but also in the role of judge, illustrates that the criminal case as a legal form is inseparable from the figure of the injured party demanding 'satisfaction', and accordingly from the more general form of the legal transaction. The public prosecutor demands, as befits a 'party', a 'high' price, that is to say a severe sentence. The offender pleads for leniency, for a 'discount', and the court passes sentence 'in equity'. If one were to reject this form of transaction completely, one would deprive criminal proceedings entirely of their 'juridical soul'. Imagine for a moment that the court was really concerned only with considering ways in which the living conditions of the accused could be so changed that either he was improved, or society was protected from him – and the whole meaning of the term 'punishment' evaporates at once. This does not mean that all of penal-judicial and executive procedure is totally devoid of the above-mentioned simple and understandable elements; we merely wish to demonstrate that this procedure contains particular features which are not fully dealt with by clear and simple considerations of social purpose, but represent an irrational, mystified, absurd element. We wish also to demonstrate that it is precisely this which is the specifically legal element.

---

[18] Nowadays, the injured party's satisfaction is regarded as one of the purposes of punishment. (Cf. Franz Eduard von List, *Lehrbuch des leutschen Strafrechts*, 1905, section 15).

There appears to be a further difficulty in the following. Archaic criminal law knew only the concept of injury. Guilt and blame, which are given such a prominent position in modern criminal law, were totally absent at this stage of development. The deliberate act, the negligent act, and the accidental act were judged only by their consequences. The customs of the Salic Franks and of the present-day Ossetians are, in this respect, on the same level of development. For example, the latter do not differentiate between death caused by a stab with a dagger and death occurring as the result of a stone, knocked by a bull's foot, rolling down the mountain.[19]

As we can see, it does not in any way follow from this that the concept of liability as such was unknown to ancient law. It was simply differently determined. In modern criminal law, the concept of strictly personal liability corresponds to the radical individualism of bourgeois society. In contrast to this, the law of antiquity was permeated by the principle of collective liability. Children were punished for the sins of their fathers, and the gens was liable for every member of the gens-community. Bourgeois society undermines all hitherto prevailing primitive and organic bonds between individuals. It proclaims the principle of 'every man for himself' and follows it with absolute consistency in all spheres, including criminal law. Furthermore, modern criminal law has introduced a psychological element into the concept of liability, thereby making it flexible; it has broken it down into degrees: liability for a foreseen outcome (premeditation), and liability for an outcome not foreseen, but foreseeable (negligence). Finally, it postulated the concept of not being answerable for one's actions (*Unzurechnungsfähighkeit*), that is, complete absence of any liability. It goes without saying that the introduction of the psychological element into the concept of liability implied the rationalisation of the fight against crime. A theory of prevention in general and in particular could be postulated only on the basis of differentiat-

---

[19] If an animal from a herd of sheep, cattle, or horses – so it says in a description of the customs of the Ossetians – knocks a stone down from the mountain, and this stone injures or kills someone passing by, then the relatives of the injured or dead person pursue the owner of the animal with their blood vengeance, or demand blood money from him, just as though it were a premeditated act of murder (Cf. M. Kovalevsky, *Modern Custom and Ancient Law*, op. cit., vol. II, p. 105).

ing between being responsible or not being responsible for one's actions. Nevertheless, to the extent that the relationship between the offender and the sentencing power was construed as a legal relation and took the form of a judicial trial, this new element did not in any way eliminate the principle of equivalent retaliation. On the contrary, it created a new basis for its application. What does this breakdown mean, if not the differentiation of conditions to be applied in a future judicial transaction? Shading responsibility is a basis for shading the punishment: it is a new, and if you like, an ideal or psychological element which is taken into account, together with the material aspect of the damage and the objective aspect of the deed, to provide, collectively, the basis for determining the degree of punishment. For a premeditated act – the gravest responsibility, therefore, under otherwise unchanged circumstances, also the most severe punishment; negligent action – a less grave responsibility, *ceteris paribus* a lesser sentence; finally, no responsibility (the perpetrator cannot be held to account for his actions) – punishment waived.

If, in place of the punishment, we substitute treatment (*Behandlung*), that is to say a concept of medical-health, what follows is entirely different, since we would then be interested, not primarily in whether the punishment fits the crime, but in whether the measures taken are adequate to the goals set, that is, whether they are adequate to the protection of society, to having an effect on the offender, and so forth. From this point of view, it can easily be the case that the relationship is a completely inverse one, in other words, that the most intensive and most prolonged measures of influence are required, precisely where there is a case of diminished responsibility.

If the punishment functions as a settlement of accounts, the notion of responsibility is indispensable. The offender answers for his offence with his freedom, in fact with a portion of his freedom corresponding to the gravity of his action. This conception of liability would be quite superfluous in a situation where punishment has lost the character of an equivalent. Were there really no trace of the principle of equivalence remaining, then punishment would entirely cease to be punishment in the juridical sense of the word.

The juridical concept of guilt is not a scientific concept, since it leads directly to the contradictions of indeterminism. From the standpoint of linking the causes which bring about some event, there is no basis for giving prominence to one link in the chain over another. The actions of a psychically abnormal person (not answerable for his actions) are no less determined by a series of causes (heredity, living conditions, environment, and so on) than are the actions of a completely normal (fully accountable) person. It is interesting that punishment applied as a paedagogical measure (that is to say outside the notion of equivalence) is in no way connected with the conception of responsibility, freedom of choice and the like, nor does it have need of this conception. In paedagogy, the appropriateness of a punishment (of course we are speaking here of appropriateness in the most general sense, regardless of the form, or of the mildness or severity, of the punishment) is determined solely by the ability to gain an adequate grasp of the connection between one's own actions and their disagreeable consequences, and to retain this in one's memory. Persons whom criminal law regards as not responsible for their own actions, such as children of a very tender age, psychically abnormal people, and others, are also responsible in this sense, in that they are open to being influenced in a particular direction.[20]

In principle, punishment in keeping with the guilt represents the same form as retaliation in proportion to the injury. Its most characteristic feature is the arithmetical expression of the severity of the sentence: so and so many days, weeks, and so forth, deprivation of freedom, so and so high a fine, loss of these or those rights. Deprivation of freedom, for a period stipulated in the court sentence, is the specific form in which modern, that is to say bourgeois-capitalist, criminal law embodies the

[20] The famous psychiatrist Kraepelin points out that 'educational work among the mentally ill, as it is indeed carried out with great success, would of course be inconceivable if all the mentally ill who are inaccessible to criminal law actually lacked the freedom of self-determination in the sense implied by the legislator'. (Emil Kraepelin, *Die Abschaffung des Strafmasses*, 1880, p. 13). Naturally the author makes the reservation that one should not read him as suggesting that the mentally ill be answerable in terms of criminal law. Nevertheless, these observations demonstrate with sufficient clarity that criminal law handles the concept of responsibility (*Zurechnungsfähigkeit*) as a condition of liability for punishment, and not in the only correct sense, as defined by scientific psychology or educational theory.

principle of equivalent recompense. This form is unconsciously yet deeply linked with the conception of man in the abstract, and abstract human labour measurable in time. It is no coincidence that this form of punishment became established precisely in the nineteenth century, and was considered natural (at a time, that is, when the bourgeoisie was able to consolidate and develop to the full all its particular features). Prisons and dungeons did exist in ancient times and in the Middle Ages too, in addition to other means of physical violence. But people were usually held there until their death, or until they bought themselves free.

For it to be possible for the idea to emerge that one could make recompense for an offence with a piece of abstract freedom determined in advance, it was necessary for all concrete forms of social wealth to be reduced to the most abstract and simple form, to human labour measured in time. Here we undoubtedly have a further example of the dialectical connection between the various aspects of culture. Industrial capitalism, the declaration of human rights, the political economy of Ricardo, and the system of imprisonment for a stipulated term are phenomena peculiar to one and the same historical epoch.

Yet while the equivalence of the punishment in its crude, brutal, physical-material form (infliction of a physical injury, or exacting expiatory payment) maintains its simple meaning, accessible to everyone precisely through this crudity, it loses this clear meaning in its abstract form (deprivation of freedom for a stipulated period) even though people continue to describe the degree of punishment here too as *proportionate to* the gravity of the deed.

It is for this reason also that many criminologists, particularly those who consider themselves to be progressive, are quite naturally at pains to eliminate altogether this element of equivalence – which has patently become absurd – and to concentrate their attention on the rational goals of the punishment. The error these progressive criminologists make is that when they criticise the would-be absolute theories of punishment, they think they are confronted only with mistaken views, confused thinking, which could be countered by theoretical critique alone. In reality, however, this absurd equivalent form results, not from the aberrations of individual criminologists, but from the material relations of the society based on commodity-production which nourish it. The contradiction between the rational purpose of protecting society or of reforming the criminal, and the principle of equivalent recompense, does not exist in books and theories alone, but in life itself, in judicial practice, in the very structure of society; just as the contradiction between the fact that people in general are bound together by their labour, and that the absurb form of expression of this bond, the value of commodities, is to be found, not in books and theories, but in social practice itself. To substantiate this we have only to consider at length a few aspects.

If, in social life, the *purpose* of punishment was really the only consideration, then the execution of sentence and, more particularly, its outcome, must excite the keenest interest. Yet it cannot be denied that, in the overwhelming majority of cases, the main emphasis in a criminal trial is to be found in the courtroom at the moment of the verdict. The interest shown in the protracted methods of influencing the offender is negligible, compared with the interest aroused by the telling moment of the verdict and the determination of the 'degree of punishment'. Only a small circle of professionally interested people are concerned about questions of prison reform; the focus of attention as far as the public is concerned is whether the sentence corresponds to the gravity of the offence. If opinion in general is satisfied that the court has determined the equivalent correctly, then everything is, so to speak, settled, and the subsequent fate of the offender is of further interest to practically no one. 'The execution of sentence' – thus bemoans Krohne, one of the best-known specialists in this field – 'is the problem child of criminal law'; in other words, it is relatively neglected. He goes on:

> Even if you have the best law, the best judge, the best verdict, but the official responsible for the execution of sentence is incompetent, then you might as well throw the law into the waste-paper basket and burn the verdict.[21]

[21] Krohne, quoted from: Gustav Aschaffenburg, *Das Verbrechen und seine Bekämpfung: Kriminalpsychologie für Mediziner, Juristen und Soziologen, ein Beitrag zur Reform der Gesetzgebung*, Heidelberg (1903), 1906 ed, p. 216; translated by A. Albrecht, as *Crime and its Repression*, with an editorial preface by M. Parmelee and an introduction by A. C.

But the dominance of the principle of equivalent retribution is not only expressed in this distribution of public interest. It is manifested in a similarly crude manner in judicial practice itself. There could be no other conceivable basis for the verdicts which Aschaffenburg cites in his book, *Crime and its Prevention*. We shall here select but two examples from a whole series: a relapsed offender with twenty-two prior convictions for forgery, theft, fraud, and so on, is sentenced for the twenty-third time, this time to twenty-four hours' imprisonment for insulting an official. Another, who has spent in all thirteen years in prison and penitentiary, and has sixteen prior convictions on charges of robbery and fraud, is given a four month sentence for fraud – his seventeenth sentence. Obviously one cannot speak either of a defensive or of a reformatory function of the punishment in these cases. It is the formal principle of equivalence which triumphs here: punishment equivalent to the guilt.[22] What else could the court have done? It cannot hope to reform an inveterate habitual offender by three weeks' detention, nor yet can it lock up the person in question for life, just for insulting an official. The court has no choice but to allow the offender to pay for his petty crime in small change (so many weeks' loss of freedom). For the rest, bourgeois administration of the law sees to it that the transaction with the offender should be concluded according to all the rules of the game; in other words, anyone can check and satisfy themselves that the payment was equitably determined (public nature of court proceedings), the offender can bargain for his liberty without hindrance (adversary form of the trial) and can avail himself of the services of an experienced court broker to this end (admission of counsel for the defence), and so on. In a word, the state's relations with the offender remain throughout well within the framework of fair trading. In this precisely lie the so-called guarantees of criminal proceedings.

The offender must therefore know in advance *what he is up for*, and what is coming to him: *nullum crimen, nulla poena*

Train, Boston: Little, Brown & Co., 1913 (Modern Criminal Science Series, vol. 6). [Transl.]

[22] This nonsense is nothing but the triumph of the legal idea, for law is precisely the application of an equal measure, and nothing more.

sine lege. What is the implication of this? Is it necessary for every potential criminal to be informed in minute detail about the corrective methods which would be used on him? No; it is much simpler and more brutal. He must know what quantity of his freedom he will have to pay as a result of the transaction concluded before the court. He must know in advance the conditions under which payment will be demanded of him. That is the import of criminal codes and criminal procedures.

One should not make the mistake of imagining the matter to be such that whereas the mistaken theory of retribution dominated criminal law in former times, it was later superseded by the correct viewpoint of social protection. It would be wrong to consider the development as if it were taking place on the plane of ideas alone. In reality, sentencing policy contained social elements both before and after the emergence of the sociological and anthropological tendencies in criminology. Alongside these, however, it contained and still contains elements which do not arise from this *technical* purpose and do not, therefore, allow criminal procedure to be expressed *absolutely and completely* in the rational, non-mystified form of social-technical regulations. These elements, whose origin should be sought, not in sentencing policy as such, but much deeper, give the juridical abstractions of crime and punishment their reality, and ensure them their practical significance in the framework of bourgeois society, despite all the efforts of the theoretical critique.

At the Hamburg Congress of Criminologists in 1905, van Hamel, a reputable representative of the sociological school, declared that the main obstacles to modern criminology were the three concepts of guilt, crime and punishment. If we freed ourselves of these concepts, he added, everything would be better. One might respond to this that the forms of bourgeois consciousness cannot be eliminated by a critique in terms of ideas alone, for they form a united whole with the material relations of which they are the expression. The only way to dissipate these appearances which have become reality is by overcoming the corresponding relations in practice, that is by realising socialism through the revolutionary struggle of the proletariat.

It is not enough to label the concept of guilt as a prejudice, in order to be able to introduce immediately in practice a sentencing policy which actually makes this concept superfluous. So long as the commodity form and the resultant legal form continue to make their mark on society, so in judicial practice too the essentially absurd idea (absurd, that is, from the non-juridical viewpoint) that the gravity of each crime can be weighed on some scale and expressed in months or years of prison detention will retain its force and its meaning in real terms.

One can, of course, refrain from proclaiming this idea in such a brutally offensive formulation, but that does not yet imply that one thereby escapes its influence entirely in practice. Changing the terminology does not alter anything in the substance of the matter. The People's Commissariat for Justice in the Federated Russian Socialist Soviet Republics published guidelines for criminal law as early as 1919, in which the principle of guilt as the basis of punishment is rejected and the punishment itself is characterised, not as retribution for guilt, but solely as a protective measure. The 1922 penal code of the RSFSR likewise does without the concept of guilt. Finally, the 'Principles of Penal Legislature of the Union' passed by the Central Executive Committee of the Soviet Union eliminate the term 'punishment' altogether, replacing it with the term 'judicial-corrective measures of social defence'.

Such an alteration in the terminology undoubtedly has a certain demonstrative value. Nevertheless, declarations do not decide the merits of the case. Transforming punishment from retribution into a measure of expediency for the protection of society and into the reform of individuals who are a threat to society calls for the completion of a colossal organisational task. Not only does this task lie beyond the sphere of purely judicial activity, but it would actually, in the event of its being completed successfully, render the court case and court verdict totally superfluous. For when this task has been fully accomplished, the reforming influence will no longer be a simple 'legal consequence' of the court's verdict, in which some 'evidence' has been deposed, but will acquire a completely independent social function of a medical-educational nature. There is no doubt that our development is proceeding, and will continue to proceed, in that direction. For the present, however, so long as we continue to have to lay emphasis on the word 'judicial' when speaking of measures of social defence, so long as the forms of the judicial process and of the material penal code are still retained, altering the terminology will remain a largely formal reform. This fact did not, of course, escape the notice of the jurists who have written about our penal code. I will cite here only a few opinions: Polyansky finds that, in the specific section of the penal code, 'the negation of the concept of guilt is purely superficial' and that:

> the question of guilt and of the degree of guilt is thrown up daily in the present-day practice of our courts.[23]

Isayev says that the concept of guilt

> is not alien to the 1922 penal code: for since it differentiates premeditation from negligence by contrasting these two cases with one another, it also distinguishes between punishment and social protective measures in the more narrow sense.[24]

Of course the fact is that both the penal code in itself, and the judicial process for which it is designed, are permeated by the juridical principle of equivalent recompense. What else is the general section of every penal code (including ours) with its concepts of aiding and abetting, complicity, attempt, preparation, and so forth, if not a method of weighing guilt more accurately? What should the distinction between premeditation and negligence be but a delineation of degrees of guilt? What is the point of the concept of not being responsible for one's actions if the concept of guilt does not exist? And finally, why is there a need for the whole particular section if it is merely a matter of social (class-based) protective measures?

A consistent realisation of the principle of social protection would not require the establishment of particular *evidence*

[23] Nikolay Nikolayevich Polyansky, 'The Penal Code of the RSFSR and the Draft German Penal Code', in: *Pravo i Zhizn'*, 1922, no. 3.
[24] M. M. Isayev, 'The Penal Code of 1st June, 1922', in: *Sovietskoe pravo*, 1922, no. 2. Cf. also Trakhterov, 'The Formulation of Lack of Responsibility in the Penal Code of the Ukrainian Socialist Soviet Republic', in: *Vestnik Sovietskoy Yustitsii* the organ of the Ukrainian People's Commissariat for Justice, 1923, no. 5.

(with which the *degree of punishment* determined by law or by the court is logically connected), but would require an exact description of the *symptoms* characterising the socially dangerous situation, and an exact elaboration of the *methods* to be used in each individual case in order to avert the danger to society. The salient point is not – as many people think – merely that the social-protective measure is linked in its application with subjective elements (the nature and extent of the risk to society), while the punishment is based on an objective element, that is on concrete evidence as defined in the particular section of the penal code.[25] The salient point is the nature of this connection. It is difficult for the punishment to free itself from its objective basis, because it cannot throw off the form of equivalence without losing its fundamental characteristic. Yet only concrete evidence provides any likeness to a quantifiable entity and is thus a certain sort of equivalent. A person can be forced to *expiate* a certain action, but there is no sense in forcing him to atone for the fact that society considers him, the subject in question, to be dangerous. That is why punishment is predicated on precisely established evidence, which the social protective measure does not require. Compulsory atonement is juridical coercion addressed to the subject from within the formal framework of the trial, the verdict, and the execution of sentence. Coercion as a protective measure is an act of pure expediency, and as such, can be governed by technical regulations. These regulations can be more or less complex, according to whether the purpose is the mechanical elimination of the dangerous individual, or his reform. In either case, however, these rules express in clear and simple terms the goal society has set itself. In the legal norms which prescribe particular punishments for particular crimes on the other hand, this social purpose appears in masked form. The individual who is subjected to influence is put in the position of a debtor settling his debt. It is no accident that the word 'execution' is used for the compulsory fulfilment of private law obligations, as well as for disciplinary penalties. The term 'to serve a sentence' also expresses the same thing exactly. The offender who has served his sentence returns to his point of departure, that is to his

[25] Cf. Piontkovsky, 'Measures of Social Defence and the Penal Code', in: *Sovietskoe pravo*, 1923, nos. 3-6.

isolated existence in society, to the 'freedom' to enter into obligations and to perpetrate punishable deeds.

Criminal law, like law in general, is a form of intercourse between isolated egoistic subjects, the bearers of autonomous private interests, or ideal property-owners. The more wary of the bourgeois criminologists are not at all ignorant of this connection between criminal law and the legal form in general, in other words of the fundamental conditions without which a society of commodity producers is inconceivable. That is why their very reasonable response to the demands of the extreme representatives of the sociological and anthropological schools – that they should lay the concept of crime and guilt to rest, and that they should put an end to the juridical elaboration of criminal law altogether – is the question: in that case, what about the principle of civil liberty, the guarantees of procedural regularity, the principle of *nullum crimen sine lege*, and so on?

That is exactly the position taken by Tzhubinsky in his polemic against Ferri and Dorado, among others. I quote a characteristic passage:

> While giving his (Dorado's) glorious belief in the unlimited power of scholarship full credit, we prefer to remain on firm ground, that is to reckon with the lessons of history and the actual facts of reality; in so doing we shall be forced to admit that it is not 'enlightened and rational' free will which is required (with guarantees that this will indeed be enlightened and rational), but a firm legal system, for whose maintenance *juridical* study will continue to be necessary.[26]

The concepts of crime and punishment are, as is clear from what has been said already, necessary determinants of the legal form, from which people will be able to liberate themselves only after the legal superstructure itself has begun to wither away. And when we begin to overcome and to do without these concepts in reality, rather than merely in declarations, that will be the surest sign that the narrow horizon of bourgeois law is finally opening up before us.

[26] Cf. M. Tzhubinsky, *Course in Penal Policy* (*Kurs ugolovnogo prava*), 1909, pp. 20-33.

Appendix:
An Assessment by Karl Korsch

[The following is extracted from an article which appeared in a reprint of the *Archiv für die Geschichte des Sozialismus und der Arbeiterbewegung* of 1930 (15th year) by Carl Grünberg. Leipzig. The reprint appeared with Grazer Verlagsanstalt in cooperation with the Limmat Veralag, Zürich. Since Pashukanis' work fell into oblivion so quickly, there has been no continuous debate on its theses. Hence we consider it worthwhile to recover for the present discussion this early reaction by Korsch. – *Ed.*]

The whole of the 'Critique of the Fundamental Juridical Concepts' and resulting 'General Theory of Law' by the Soviet Marxist jurist Pashukanis consist solely of positing and rigorously developing the formula that not only the varying contents of currently obtaining legal relations and legal norms, but also the legal form itself in all its manifestations is 'just as' fetishistic in character as is the commodity form of political economy. Like the latter, law in its fully developed shape pertains exclusively to the historical epoch of capitalist commodity production, and in a parallel historical development, had insignificant beginnings, only becoming recognisable in the course of subsequent development. In the bourgeois 'constitutional state (*Rechtsstaat*)' of today, law has spread, in part actually, in part potentially, from its original sphere of regulating the exchange of commodities of equal value to affect absolutely all social relations existing within modern capitalist society and the state governing it. Together with capitalist commodity production, the bourgeois 'state', its classes and class antagonisms, law will be completely transformed in its content in the communist society of the future, and will ultimately also 'wither away' altogether as a form.

It is obvious that a critique of the historical phenomenon of law in its entirety which starts out from so radical a materialist principle and goes right to the heart of the matter must lead, when carried through consistently, to extremely far-reaching consequences, with many of the conceptions which up to now even the socialist critics accepted more or less unquestioningly, being turned upside down. This revolutionary theoretical significance of the book to hand is not in itself adversely affected by the fact that all these revolutionary ideas put forward by Pashukanis are not actually new, but restore and renew the same ideas expressed by Marx himself, partly by implication, but to a large extent explicitly as well, as many as eighty years ago in his critique of *German Ideology*, in the *Communist Manifesto*, and repeated some decades later in *Capital* and the *Critique of the Gotha Programme* (1875). For between the two lies a long period in history in which these consequences of the original revolutionary Marxist theory were as totally forgotten in the sphere of law as they were in the sphere of politics. The radical Marxist tendency in the West and in the East was only able to unearth these consequences from the oblivion in which they had languished for decades, in the new, critical, period of capitalist development after the turn of the century, and in the sharpened class struggles of the War and post-War period. Only then could they be restored to their pure form by the removal of the reformist and bourgeois distortions which had accreted to them in the intervening period.

For this reason, it does not seem to us to be especially vital to the critique of Pashukanis' theoretical achievement in rehabilitating the Marxist theory of law that, despite the 'orthodoxy' to which the author aspires, he has not in fact in his book restored all the consequences of Marxist theory for the sphere of law with absolute consistency; indeed he has not even re-stated all those already clearly expressed by Marx himself. Rather, despite his forceful beginning, he still evades in the end some of the most far-reaching and bold consequences.

In his penultimate chapter for example, he shies away from the conclusion which his portrayal, accurate enough in itself, of the connection between law and morality in the commodity-producing class society of today has led him to. That conclusion should have been that, after the proletarian revolution has been fully accomplished, after commodity production, classes and class antagonisms have been abolished, and after the state and law have 'withered away' completely, 'there will be no morality' any longer in the communist society of the future, freely developed on its own basis. But in a footnote, added explicitly for this purpose, he restricts the 'withering away' in this sphere to the 'specific forms' of moral consciousness and moral conduct peculiar to the present historical epoch which, once they have played out their historical role, will make way for 'different, higher forms'.[1]

And in another place, in analysing the problems of 'law and law-breaking' in the final chapter of the work, he even goes so far as to make explicit mention of a new 'system of penal policy' to be created after the complete disappearance of classes. For while he does raise the question as to whether 'in such circumstances, there will be any necessity at all for a penal system', he obviously confines his own perspective to the abolition of the 'legal fabric' and the 'characterisation "criminal law" '.[2] In contrast, as early as the *Communist Manifesto* of 1847/48, Marx and Engels listed explicitly, amongst the most general forms of consciousness which, despite all variety and differences are common to all the centuries of the history of class society until today, and which will 'completely vanish with the total disappearance of class antagonisms' in the epoch of the proletarian revolution – besides 'religion', 'philosophy', and 'politics' – unconditionally and without any reservations whatsoever, 'morality' and 'law' in their entirety. Marx and Engels explicitly rejected the mere 're-shaping' of their previous forms.[3]

We are far from reproaching the 'orthodox Marxist' Pashukanis with these and several other instances where his critic-

---

[1] E. Pashukanis, *The General Theory of Law and Marxism: Towards a Critique of the Fundamental Juridical Concepts*; (in this edition: chapter 6, footnote 12, p. 160).
[2] Ibid., p. 176.
[3] Marx and Engels, 'Manifesto of the Communist Party', in: Marx and Engels, *Selected Works*, Vol. I, 1969, pp. 125-126.

ally revolutionary 'theory' lags behind the theoretical ideas expressed by Marx and Engeles themselves in an earlier historical period. We tend rather to see the decisive failing of this 'materialist' critique of law in its all-too didactic, scholarly-dogmatic character, which, compared with past and present reality and practice, is frankly 'juridically unworldly' at a theoretical level. In this connection, it is particularly illuminating to compare this 'general theory of law' published by the Russian Soviet Marxist Pashukanis in 1929 [Korsch is mistaken here; while the German translation from which he is evidently working was published in 1929, the original work was first published in 1924, Transl.], not with Marx and Engels' earlier views, originating in time and in their subject-matter from the conditions of the revolutionary dawn of the working class movement, but with a work by Engels (together with Kautsky) which appeared in *Neue Zeit* in 1887. Engels here expresses his views on questions of law with reference to the practical and theoretical demands of a new stage of development in the modern working class movement much more closely related to present-day conditions. Sharply as the materialist critic Friedrich Engels takes issue in this piece with the illusions of the 'jurists' socialism' put into circulation at that time by Anton Menger and other 'well-meaning' friends of the workers; forcefully as he emphasises that the modern working class 'cannot fully express their conditions of life in the juridical illusion of the bourgeoisie'; cuttingly as he rejects the implication that one of the existing socialist parties would dream of 'turning their programme into a new philosophy of law'; still less is he simply content with this negation – abstractly attuned to the revolutionary 'ultimate victory' – of the 'legal form' and the 'juridical attitude' which essentially pertain to bourgeois society. Rather, he contrasts Menger's ambitiously posited (but theoretically inadequate and ultimately unattainable) socialist 'basic rights' of workers to the 'full proceeds of their labour', with the different 'specific legal claims' which he feels socialists should be raising, and without which 'an active socialist party, and indeed any political party at all, is impossible'. And the only fundamental condition he sets for such a programme of legal claims by the struggling proletarian class consists of the materialist condition that all these claims, varied and variable depending on period, country, and level of social development, must in all cases be precisely adapted to the relations and conditions of the class struggle.

It is quite evident that the standards set in this testament by Friedrich Engels for the evaluation of a socialist legal programme and hence also of a socialist theory of law cannot simply be applied without alteration to the 'Marxist theory of law' of the Soviet Marxist Pashukanis. Pashukanis' book was predicated on the totally different situation of a proletarian revolution already in process, and even those who regard this predicate as an historical illusion and a delusion must take its subjective existence into account in evaluating the theoretical content of this 'Marxist theory of law'. One cannot even reproach the author with having ignored the bourgeois character of existing institutions in his own field, that of law, in his 'socialist Soviet state' in the present 'transition period', a character unaltered by changes in name. On the subject of the criminal law currently valid in the Soviet Union, which erased the concept of 'guilt' from its statutes as long ago as 1919 and 1922 (meanwhile retaining such forms of guilt as 'intent', 'negligence', and the fundamental concept of legal guilt, 'irresponsibility' – *Unzurechnungsfähigkeit*), and which some time ago replaced the concept of 'punishment' with the characterisation 'judicial-corrective measures of social defence', he states with refreshing emphasis that thus 'changing the terminology does not alter anything in the substance of the matter'.[4] Nevertheless, the sole fact that the Soviet author naturally hangs onto the concept of the 'transition period', and basically sees the whole development occurring in Russia at the present time, in the legal, as well as in the political, economic, and every other sphere of life, as an evolutionary transition to communist society after the complete overthrow of the capitalist social order, means that his whole approach is inevitably illusory, since it attempts to comprehend current relations and developmental tendencies, not in materialist terms, according to their concrete nature, but idealistically, in the light of a subjectively posed goal. The extraordinarily abstract nature of this work, to which we have already drawn

[4] Pashukanis, *The General Theory of Law* . . .; (this ed.: p. 185).

attention, and which in parts intensifies to become downright scholasticism, arises in the last analysis from this, and not from coincidental causes such as that this work was originally intended as a first draft, written largely as a means for the author to clarify his own ideas.[5]

This outmoded scholasticism of method, which cannot be overcome by theoretical means in the given circumstances, is employed by Pashukanis in his internally inconsistent attempt to rehabilitate the pure and unfalsified revolutionary Marxist critique of law as the theoretical expression of the actual historical development presently occurring in the Soviet Union and, indirectly, on a world scale. And this method ultimately and inevitably leads him to distort, theoretically, that very theory whose 'pure and unfalsified' rehabilitation he is subjectively striving to accomplish to the letter.

Marx and Engels make a fundamental distinction between the inherently 'fetishistic nature' of the commodity form on the one hand, and the higher 'ideologies' based on this form – politics, law, the ideologies of philosophy and religion and the like, which 'distance themselves even further from the material economic base' and are in this sense 'even higher' ideologies – on the other. Contrary to this view, Pashukanis' entire 'Marxist' critique of law and 'general theory of law' rests on the consistent equation, whilst not of law and the economy, yet of the legal form and the commodity form. The entire colossal process of development in actual history which led to the emergence of the Marxist materialist view of law, the state, society, and history, and to its critique of political economy, a materialist view which is still preserved in its consummate shape, is as it were absolutely obliterated and even partially revoked by Pashukanis' method. When he speaks explicitly of two equally 'fundamental' aspects of the homogeneously integrated relation of people living in commodity-producing society, an economic and a legal aspect, when he explicitly characterises 'legal fetishism' and 'commodity fetishism' as two equally 'enigmatic' phenomena resting 'on the same basis', when he says that 'both these basic forms' are 'interdependent' and that the social link rooted in production is

---

[5] Ibid., preface to 2nd Russian edition (here: p. 37).

manifested in these 'two absurd forms' simultaneously,[6] Pashukanis is deviating decisively, as he does in countless other instances which permeate his book as a homogeneous thread, from the Marxist idea which regards the economic relation as the fundamental one, with the legal relation on the contrary, like the political relation, as derived from it. Let us consider in addition his polemic – which in itself is accurate, but overshoots its target – against Marxist critics of law like Reisner, who wish to conceive of law purely and simply as an 'ideology' rather than as the expression of an actual social relation, albeit ideologically masked and distorted. Let us consider his equally accented opposition to all the older as well as the more recent socialist and communist theoreticians who have seen the clarification of the class nature of the form of law in its entirety, as well as of its specific content at different times, as the most substantive issue for the Marxist critique of law. Let us consider his extremely strange – for a 'Marxist' – over-estimation of 'circulation', which he regards, not only as a basic determinant of the traditional ideology of property, but also as the only economic reality on which property is based today. And finally, let us not fail to take into account his conspicuously 'strange' attitude to economic theory and history in general. What all this adds up to is a total picture of a critique of law and a 'theory of law' which, despite its strictly materialist and 'orthodox Marxist' methodological starting point, distances itself in its actual execution and in its results from the materialist, critical, theoretical, and at the same time potentially practical, revolutionary destruction and abolition of legal ideology and of the economic social reality of capitalist society on which it is based. What he tends towards is, at the level of theory, a renewed recognition and re-establishment of legal ideology and the reality it masks. In the same period, actual historical development moved and is moving in the same direction. We mean the entire economic and social development occurring in the Russian Soviet Union, which has as one of its component parts the specific sphere of its legal historical development, whose ideological expression and reflection we see in the theoretical work by Pashukanis with which we have been dealing.

[6] Ibid., (here: p. 113 and p. 117).

Printed and bound by CPI Group (UK) Ltd, Croydon, CR0 4YY

02/02/2025

14636101-0004